ISBN: 979-8-9930525-0-2

Contents

Introduction

The purpose of this book is to advocate for a transformational shift and re-imagination of the way we practice philanthropy, using technology as our medium for change. The primary objective is to explore how foundations can incorporate technology into their work to enhance and scale impact. We will delve into ways technology can be seamlessly integrated into the philanthropic sector, benefiting charities, foundations, people and organizations to ultimately change the way we give.

Having worked in the nonprofit and philanthropic sector for my entire career, not much has changed. The way we operate is the same. The way we govern and finance our work is the same. While there are great examples of passionate people and societal impact, we have work to do as a sector.

Technology creates the ultimate opportunity to transform our work. But unfortunately, we spend less on technology and don't fully fund the education or training required by nonprofits to utilize technology in new ways. We spend too much time pondering whether it is ethical to use AI.

Other sectors use technology to do things better and more effectively. Even with the developments brought on by the internet, the philanthropic sector has been slow to incorporate these new technological solutions into our work. Astonishingly, according to GuideStar in 2020, only 10% of foundations even had websites.

The philanthropic sector is facing significant challenges that require urgent solutions. The problems we face today are larger and more complex than ever before, necessitating a wholesale shift rather than incremental changes. My purpose in writing this book is to lay out a vision for reimagining the way we as foundations and donors give. I will offer new ways of thinking and question some of our most basic assumptions.

We need to reimagine the philanthropic sector. If our solutions don't evolve to the scale of the current societal and global challenges, our work and efforts will never pay off. At a global scale, global warming and super storms demand urgent solutions. Locally, within our own communities, most of us reading this book know of homeless people, isolated seniors living alone, disabled individuals who struggle to access health care, veterans who can't get dental care or teenagers struggling with depression. To those of us directly affected or trying to help someone directly, these challenges may resonate more than the larger global challenges.

Reimagining our work starts with philanthropic organizations. The philanthropic sector, along with our vital nonprofit partners, is in the unique position of trying to solve these problems, but unlike the governmental sector, philanthropy can evolve and change. When it comes to producing results, philanthropy enjoys a long-time horizon, thus we have more room to innovate and explore.

This is not a book about using technology. Moving from paper to computers, or from in-person meetings to zoom, is not a reimagination of philanthropy. A truly reimaginative process requires concerted effort and an acknowledgement that we must move beyond our existing assumptions.

It is a book about using technology as the basis to reimagine our sector and ultimately change the way we give. It is about rallying our collective efforts to implement broad systemic reforms and capture opportunities for wholistic adoption of new ways of working offered by technology to truly reimagine our work. We must start preparing now for the next decade.

Although most of the technologies discussed in this book are already available, they are barely utilized nor anywhere near a state of full adoption. This book will provide key examples, takeaways, and actionable steps that leaders can take to integrate technology into their operations, thereby increasing their overall impact. We will then examine emerging technologies and concepts to reimagine the way we give.

In Part 1 of this book, we will start by defining the substantial benefits should we decide to take this journey. We will then examine some of the inherent barriers and long-held assumptions that must be collectively dismantled. Next, we will consider opportunities to reimagine our sector created by AI, Sensors, the Blockchain and Ethereum. We will then discuss some of the risks of playing a passive role in the transformation of our sector. Finally, the last part of this book will offer our strategic recommendations for foundations.

Preface

This book primarily addresses leaders, staff and board members of foundations. Foundations provide key financing to nonprofits and essentially either support or direct the trends and methods in the nonprofit sector. It also touches on other audiences that play key roles, namely donors and supporters of nonprofits. Leaders of nonprofits may also find this book useful, as it is always a good strategy for a nonprofit to know the focus of their funders. The philanthropic and nonprofit sectors (the sectors) are intertwined, interdependent and reliant on each other. Albeit the case, the bulk of this book focuses on improving philanthropy. However, as someone with experience working in both sectors, I will use the first person in this book and may address each sector separately or couple them collectively as 'the sectors.'

The work is based on my 30 years of experience working in the nonprofit and philanthropic sector, in roles ranging from a teenager raising money door to door for the Multiple Sclerosis Society, to a Peace Corps volunteer and then a C-Suite professional leading the strategy for one of the largest health foundations in Florida. In addition, I have served and continue to serve as a founder, volunteer, board member, advocate and advisory member for various nonprofits, both large and small.

Some elements of the book can be used as a model or business plan to sustain and support the reimagination of the nonprofit and philanthropic sectors. Chapters may include a brief case study. The case studies are designed to demonstrate how we can start to think differently, see the practical application of technology, and create new solutions. The case studies are based on my real-life experience working in the nonprofit and philanthropic sectors for 30 years. The last section includes sample philanthropic grantmaking strategies which can be used to guide implementation.

Finally, while it is obvious who some of the individuals are in this book, given their fondness for litigiousness, their position of authority and my lack of sponsorship or a publisher for this book that can back me, I've chosen not to name names.

PART 1

Giving Reimagined

We always overestimate the change that will occur in the next two years and underestimate the change that will occur in the next ten. Don't let yourself be lulled into inaction.

BILL GATES

What does it look like when we reimagine giving? How do we do it? What is our first step and what else is involved?

We start by incorporating best practices, techniques and trends from other industries while simultaneously re-thinking and questioning the underlying assumptions of our sector.

Incorporating existing Artificial Intelligence (AI) tools into our work, we can create robust organizational records that lend to organizational memory, record-keeping and long-term sustainability. These tools can also be used to obtain information on projects or initiatives as they develop, while allowing the creation of vast bodies of knowledge we can use to educate individuals and society.

By using the capability of AI to connect our servers and our vast troves of records, information will be shared in real-time among everyone and is not held in subscriber-only domains, nor siloed within our own server.

Leveraging AI, we can share our unique perspectives and knowledge within publicly available data platforms that update in real-time to present meaningful and useful information to philanthropists and donors. Information that is freely accessible and shared in a way that anyone and everyone can access, to allow for real-time updates, insights and decisions.

Then, using the compounding effect of Generative AI, we can take our sector specific knowledge to the next level. Foundations using connected systems leveraging Generative AI will allow us to create real-time data to glean critical programmatic insights which flow seamlessly between individuals, donors, organizations and foundations.

We can incorporate sensors and smart glasses into the donor experience and even leverage these tools to gain a whole new level of insight into whether a program works. This hardware can also reduce the time it takes to donate by connecting with new digital tools for seamless due diligence and donations. Digitally connecting interested donors with nonprofits increases accountability and allows for instant donations. When donor activity and the flow of donor capital is captured within our Generative AI systems, it builds a greater and more resilient body of knowledge around what works that we can share sector wide.

We can experiment with emerging technology, such as the blockchain, which is the underlying technology upon which Bitcoin is built. The blockchain is essentially a public ledger that records all transactions and can be viewed by the public to verify them.

Adopting the tenet of decentralized ownership and transparency, we can open a new method of governance and oversight that results in a greater ability to monitor the performance of our grantees while financing impactful projects. We will learn how the blockchain can be used to create a collective decision-making model that leads to higher levels of engagement on behalf of all participants in a project, thereby increasing transparency and eliminating the role of the foundation as administrative middleman. The system will allow us to tap into new systems of oversight and governance. Lastly, by using the blockchain, we can bypass the traditional financial system, with its fees and unwillingness to finance innovative financial vehicles such as stablecoins and real estate.

We can migrate our operations onto the shared governance internet network which is the Ethereum 'chain.' Ethereum is a digital ledger separate and independent of the existing internet and not bound by the existing constraints of our existing administrative and governance structures. It opens up entirely new worlds of collective oversight, shared contracts and payments which allow for performance-based payments that flow more consistently to nonprofits. Nonprofits implementing collective impact projects connected to the Ethereum chain are more closely engaged in the work and have greater ownership of the results. Groups of nonprofits monitor the progress of their efforts jointly, and provide verification for the work of others, thus reducing the administrative burden on the funder by eliminating the need for reporting or constant oversight by the program officer.

Reimagining our sector will allow for many advantages. We will come to enjoy and benefit from more rapid decision making by sharing data across sectors and silos. It will incorporate technology available to most of us to alleviate suffering, while preparing to use technology currently out there but not available for broad adaptation.

The results of utilizing technology to reimagine how we give are clear. The possibilities expand daily as new technology comes into play. In the end, we can create:
- Better insight into what works
- Better research and faster adoption of solutions
- More responsive programs and services
- More concise and relevant impact measures
- Increased access to new and transformative financing
- Empowered people and donors with ready access to data to drive decisions
- Better sense of the impact of our giving and the impact of the entire sector

- More real-time feedback to pinpoint what and who should be supported
- More reliable data to demonstrate our global impact
- Greater accountability and transparency for the sector
- Improved learning and increased knowledge
- Improved ability to scale solutions that work

Our first step is to acknowledge the benefits that technology offers and start exploring short-term opportunities to immediately integrate technology into our work. Then over the long term, we build new administrative, compliance, oversight and financial systems upon emerging technologies to truly transform the way we think and operate.

Admittedly, using these new methods will disrupt existing power structures, erode our core beliefs and eliminate outdated systems, structures and principles that ultimately hinder progress. To reach maximum effectiveness and create an environment ripe for innovation, truly reimagining our sector will require wholesale re-writing of the administrative, policy, privacy, governance and data usage policies we currently employ. Nonetheless, if collectively developed and implemented, our transformation can lead to more rapid decision-making, more efficient workflows, targeted interventions, improved outcomes at scale and stronger, more transparent operations.

Merriam-Webster defines a transformation as the 'act, process or instance of transforming or being transformed.' Transformations change our mindset, operating model and eventually the way we conduct business. Transformations happen to someone or something, like the caterpillar transforming into a butterfly. As we will highlight in this book, a digital transformation is already underway in our sector.

Conversely, the act of reimagining, while also likely transformational, is a process we undertake ourselves. It is not something that happens to us, as does a transformation. Reimagining starts with us. As we have seen the spark of technology transform other sectors, inevitably, technology will transform philanthropy and afford us the opportunity to reimagine the way we give.

It is incumbent upon us to leverage the power of technology, take ownership of this change and reimagine our sector. Taking ownership of this process will allow us to shape the ultimate outcome. However, where this leads or what opportunities this creates, is to some degree unknown. But such are the most intriguing journeys; those that lead us down unknown paths, with potential for great discovery.

CHAPTER 1

Technology

Change is coming in the technology sector. <u>Eric Schmidt, the former CEO of Google, in an interview with Tim Ferriss</u>, when asked about upcoming advances in the tech sector said, "I guess for me, it's software and analytical thinking. I am a believer that the next 50 years, human society will have incredibly complicated human systems. So if you think about the things we deal with every day, the judicial system, the political system, the prison system, the traffic system, what have you, they were architected in a world where we didn't have a lot of data, and we didn't have a lot of software, and we couldn't really measure everything. And I think a lot of those systems are going to get very, very thoroughly designed. And if you're going to design those systems, then design based on outcomes you care about.'

The <u>United Nations Secretary General's Roadmap for Digital Cooperation</u> starts with the acknowledgement that 'The world is shifting from analog to digital faster than ever before, further exposing us to the vast promise and peril of new technologies. While the digital era has brought society many incredible benefits, we also face many challenges such as growing digital divides, cyber threats, and human rights violations online.'

According to a recent article in the Economist, AI will have a 'compounding effect' on the economy. They further point out that 'the progress of AI has for the best part of a decade outpaced forecasts of when it would pass various benchmarks.' The article concludes by saying that even if our predictions still fall short of the promised capability, we are in for a 'big surprise.'

In the tech industry, there is no homeostasis. Every day someone, somewhere, wakes up thinking about how to improve existing software, hardware and technological tools. The evolution of technology is constant.

This principle of technological advancement and the doubling of transistor speed is called Moore's law. Traditionally, technological capability doubles every two years, which could even be faster now with Quantum computing and other advancements. But even that assumption could be up for debate in the future given the technological evolutions and revolutions happening around the world today.

Since the advent of the internet, laptops and cell phones, technology has brought us many things and greatly improved our capability to accomplish a wide range of tasks. Getting around is easier, purchasing a product is easier, communicating with others is easier. Not to mention faster and more efficient. The advances in productivity and myriad possibilities technology generates are fascinating. Even more fascinating is the unknown benefits we have yet to harness.

Technology has and will transform entire fields, industries and sectors. Technology is transforming other fields: law enforcement, finance, transportation. This is called a digital transformation. Who thought a doctor could monitor your blood pressure by accessing data from your bathroom mirror? Who thought we could regulate nicotine intake to help a smoker quit smoking by applying a patch onto our arm? Moving forward, should the philanthropic sector rely on the same systems, laws, techniques, methods that we have always used? Considering the possibilities technology holds, the opportunity to create a new vision for the field and reimagine our work is upon us.

Computer hardware and software are incredibly advanced and changing daily. We have devices we can wear that tell us how long we slept and whether we are stressed at any given point during the day.

There are satellites circling the planet that can provide geo-located data. We have new digital networks and payment solutions built on top of the internet. If harnessed and nurtured, this technology can be transformative. The rate of change and adoption of new technology including artificial intelligence, the blockchain and other hardware means we don't even know where these innovations will lead us. Nor do we know what benefits will emerge from their use. Nonetheless, the time to prepare and position the nonprofit and philanthropic sectors is now.

CHAPTER 2

Political Reality

Before proceeding into this book, let's briefly set aside our goal of reimagining the way we give for a timely reflection on our current political reality. The purpose of this section is not to be alarmist.

Granted, when I first started writing this book in 2018, the political reality was a concern for the sector. Events outside our purview captured our attention and provided us with thematic and strategic rallying points. We won't dwell too much our current reality, but it can't be ignored. Furthermore, it only lends to further supporting the notion put forth in this book.

Reflect back to 1952, when, in a McCarthy-esque move, philanthropic leaders from the largest foundations were hauled before Congress to defend accusations concerning their giving. Leaders from the Carnegie Foundation were in the hot seat during these hearings and were called to attest to the perceived slights of the sector. Nothing really became of the hearings, but one result was the creation of the 'glass pockets' movement. Started by Carnegie Foundation after the hearings, this was an effort to increase transparency in philanthropy. The foundation held steadfast to this principle and as recently as 2016 stated that 'we believe that in order to earn the public's confidence, we owe the American people transparency in all of our financial and programmatic activities.'

Glass Pockets became a sector-wide movement that aimed to create more transparency around giving while also alleviating the burden on grantees. GuideStar led this effort until 2022 when the movement fizzled due to a lack of participation on behalf of foundations. Essentially, it became difficult for GuideStar to maintain the requisite digital connections among funders that a sector-wide movement toward transparency required. In their requiem for the movement, they cited the fact that only 10% of foundations even had websites in 2020 as the most obvious barrier, coupled with a lack of participation by foundations and the movement.

It doesn't take much of a stretch of imagination to imagine this same type of hearing today. Today, exorbitant fines and accusations toward large institutions that rely on philanthropy and government funding are bandied about freely at the highest levels of our government.

Unlike others, because of a lack of transparency, an accountability to no one and not to mention the enormous pool of capital retained by our foundations and endowments, our sector almost begs to be scrutinized, nay turned upside down. It is increasingly under the microscope.

In the summer of 2025, Harvard University came out publicly to say they are considering spending $500 million to settle disputes with the federal government. A recent article in the New York Times pointed out that 'neither Harvard nor the government has publicly detailed…what allegations the money would be intended to resolve.' $500 million to settle a dispute, the basis of which neither side will explain?

There was a time when such a substantial fine would likely have been discussed and thoroughly litigated. Seems that time, at least for now, is past. The sector is experiencing a world where political acts of retribution wholly dislocated from any basis in reality are the accepted norm. Rhetoric and recent political developments should create a sense of urgency around the reimagination of our work.

Government officials have adopted increasingly hostile rhetoric as an excuse to claw-back, shut-down or cut funding for nonprofits. Elon Musk, as vocal leader of the Department of Governmental Efficiency (DOGE), stated on the Joe Rogan podcast in 2025 that the nonprofit sector 'is a gigantic scam…like one of the biggest, maybe the biggest scams ever.' These are troubling words coming from the highest level of government.

Issuing fines on university endowments, drastic pullbacks if not outright elimination of federal agencies and funding are the norm of late. With no basis in fact or evidence, the tone and actions around these efforts feel all too retributive. My prediction is that we can no longer take it for granted that our work will get a pass because of how many wonderful things we do and how many lives we change.

Currently, with our push to cut taxes and reduce spending, foundations and endowments are increasingly seen as a source of tax revenue to offset other tax cuts. Without a unified collective message around the impact of our work, we are seen as faceless groups with questionable use-cases. Those who work in the sector know this isn't the case.

We should re-frame our thinking around these recent political developments and view them as an opportunity to take ownership over the future of our sector. Proactively push for transformation or wait passively and let someone else dictate the terms of how we operate and what we can and cannot do, the decision is ours to make.

PART 2

Essential Elements

Before we move into the discussion around how we can utilize technology to reimagine our sector, let us discuss essential principles of this effort. We will apply these principles in later sections of the book, and each will inform us about this process of reimagining our sectors. In terms of specific technology, we will explore that in later chapters and apply the principles covered below into those sections.

CHAPTER 3

Technical vs. Adaptive Problems

Technical problems are straightforward problems with well-defined solutions, often requiring expertise or existing knowledge to solve. Think of the hammer and nail. How do you embed a nail into a piece of wood? You hit it with a hammer. Hunger can be a technical problem. How do you solve hunger? You provide more food.

Nonprofits and foundations seek to solve both technical and adaptive challenges. Solving technical problems is easier and only requires more resources. Solutions are generally easy to identify. We can use our existing mental, organizational and governance frameworks along with verified traditional approaches to solve these problems. But that isn't the case with adaptive problems.

Adaptive problems are complex, ambiguous, and often require learning and adjustments to cultures or beliefs. The person or organization who labors to solve an adaptive challenge must be nimble, patient, innovative all while fostering visionary leadership. How do you solve world hunger? By providing food of course but there are a myriad of other elements and parameters, cultures and beliefs that must be navigated. When you zoom out and assess the problem from a macro-scale, it's not simply 'provide more food.'

Adaptive problems require new approaches and new ways of thinking called adaptive solutions. They require the courage to lay aside our existing beliefs around what is possible. In the philanthropic sector, our current system of governance and oversight is built around solving technical challenges and may not lend itself to the creativity and foresight required to adapt and evolve our work to meet the adaptive challenges of tomorrow. This requires us to start constructing a solution from an entirely different or unique angle. New frameworks and models are required. New ways of thinking are required. New tools are required. This is where technology enters the discussion.

Through the hard work of nonprofits and the generosity of foundations, we have made great advancements over the past few decades. But even if we are trying to solve a technical problem, the nonprofit sector still struggles to solve both simple and great challenges. The list of problems which require urgent solutions is long and the scale of problems is global.

CHAPTER 4

Decentralization

Philanthropic decisions are based on a centralized structure which is highly independent and comes with minimal to no accountability. Within foundations, board members or family members make decisions about who gets the money. This lack of accountability and zero transparency around decisions inherent in our centralized decision-making structure warrants further consideration, especially in a political environment of heightened skepticism. Reimagining our decision making and pivoting away from the centralized structure on which we rely toward a more decentralized decision-making structure may serve the sector well.

Recent movements in the field including collective impact and participatory grantmaking seek to build out a decentralized model of decision-making and accountability. While moving the needle to a degree, they lack true accountability mechanisms and the ability to verify the impact of our grant dollars which technology can now provide.

Chapter 5

Digital Public Infrastructure (DPI)

The Internet today consists of what could be described as 'accidental public infrastructure.' This infrastructure is comprised of social media platforms like Facebook, news sites, advertising marketplaces like Google AdSense, or music distribution networks like Spotify. These sites collectively build an 'infrastructure' which are only public insofar as they are open to the public and therefore host an increasing proportion of our civic life. They often exploit the private data of their users while restricting those users from contributing meaningfully to operations and governance.

Groups such as the Initiative for Digital Public Infrastructure at the University of Massachusetts Amherst are building a 'new, more resilient Internet for the public good; an Internet guided by the values of users and their communities rather than those of corporations and investors.' Nonprofits and foundations utilize centrally controlled websites and systems to leverage and manage volunteers. A number of sites exist which help to research potential donors. These systems require payment and may require that you provide your donor list, among other extremely sensitive information.

Foundations also utilize software systems for grants applications, grants management and grant reporting. Again, these are subscription based and the company that offers the website or service obtains all the information that goes into the system.

Most if not all our data and systems are centralized, either on our own server or on the server of the companies who manage our data.

Data is not distributed sector wide nor is it even available at scale. Yes, you can access your data and you still 'own' your data, but you have no direction in how the data is presented or how they utilize you information to inform your work. Furthermore, any data you provide is sold to another vendor or used by the company for other financial means.

Imagine systems built by and for foundations to meet our collective needs rather than those of corporations and investors. Systems that are guided by our own values and the needs of our users. Otherwise, continuing to maintain a system whereby foundations are structured and reliant on fragmented programs and software will continually detract from our ability to develop a clear picture of the value and impact of our work. Disconnected systems do not allow us to manage and utilize our data to reach our full potential.

CHAPTER 6

Case Study

During the period from 1998 to 2000, I taught English as a Second Language as a Peace Corps volunteer in Latvia. I joined the Peace Corps to travel, immerse myself in another culture and learn another language. Latvia was annexed by the Soviet Union in 1948 and lived under Soviet rule until 1991. When I arrived in Latvia, coming from the United States, I was a little surprised by the ubiquitous of cell phones. At that time back in the US, not everyone had a cell phone and we mostly used landlines. My assumption as an American was that I was coming from the most developed country on Earth, and surely this small former republic of the Soviet Union would not be using more advanced technology than we were at home.

That assumption proved false. Once I arrived, it was quite startling to see everyone walking around with cellphones to their ears. The people I met through my work volunteering with a nonprofit in Latvia staffed by college students, and even many of the other Latvian people my age (and older) were using cell phones not just to make calls but also to send text messages. The ubiquity of the cell phone and text messaging were new concepts for me. Most homes and apartments still had rotary phones but the connection was spotty and crackly.

Sometimes technology gets so old and outdated, it's use case evaporates. Infrastructure investments in the Soviet Union were managed by central planners in Moscow for decades and outlying republics (where I found myself) were low on the list in terms of priority among the central planners.

In these instances, it is better to just re-formulate a solution instead of making minor tweaks to the existing technology. Such was the case with landlines. They were so old and the hardware (rotary dial phones) was so out of date, using these phones was like going back to the late 1960's. Luckily my hearing was more robust than it is today so I could actually hear the person on the other end of the phone. In the end, it was easier and cheaper for Latvia to just migrate directly to mobile phones.

My Latvian hosts realized this and instead of waiting on the slow migration away from landlines and reliance on existing technology, Latvians took ownership of the challenge, leap-frogged the system and immediately adopted cell phones. They never looked back. For them, moving to cellular networks was not an incremental process, it was a wholesale technological transformation.

When I arrived back in the US in 2000, cell phones were more prevalent but landlines and the fax machine were still widely used. Cell phones with access to the internet and apps were another 8 years into the future.

When I got my first cell phone, I never thought that one day this same technology would be used to send money to another person via an app, or a text message. This experience with the broad adoption of the in Latvia cell phone taught me a lesson. Society can change quickly. Technology, when adopted on a mass scale, can lead to massive, unforeseen changes.

PART 3

Sector Barriers

Now that we have established the difference between technical and adaptive solutions and covered other essential themes to inform the reimagination of philanthropy, it is essential to consider some of the barriers (self-inflicted or otherwise) to reimagining philanthropy. It is important to acknowledge that many of these barriers are perpetuated by foundations.

To inject a degree of practicality into our imaginative process, let's look at some of the barriers within the philanthropic sectors that inhibit our progress. These range from administrative, financial, governing and legal.

These systems establish the framework for how we function. Nonprofits receive money from donors and foundation. Donors enjoy a tax credit for donation. Nonprofits use these funds to provide charitable services. Foundations relinquish a pre-determined amount of money annually to nonprofits, mostly based on the amount they are required to spend, which is based on a determination by the Internal Revenue Service (IRS).

Nonprofits constantly strive for innovation, partly stemming from our need to do more with less, but also given the expectations from funders to innovate. Another unique challenge to the nonprofit is the fact that they can be under continual pressure to evolve their programs and services to meet an ever-expanding need, even though they have to do this with limited resources and staff. Donors and foundations always want more innovative approaches to solving the world's intractable problems that have faced us for centuries (homelessness, hunger, environmental degradation, etc.) The same can be said of foundations, but foundations enjoy the special status of financier, thus their motivation for change is not as broad nor is there a financial motive. Nonprofits do not have the same privilege.

Let's acknowledge that even as foundations are free from the bounds of fundraising and pleasing a donor, foundations change slowly. The reason is simple: they don't need to change. Kris Putnam-Walkerly, MSW writes about this pace of decision making among foundations in her book 'Delusional Altruism.' Putnam-Walkerly breaks down the many reasons for the slow pace of decisions in philanthropy and by philanthropists, which tend to be self-inflicted barriers that philanthropists erect. These can include everything from personal reasons, over thinking things, creating too many steps to receive funds via a long and cumbersome grant application and even refusing to donate to a cause lest people view them as 'second comers' to after another donor lays their stake in the ground and makes a gift. This somewhat sclerotic process and way of operating impacts our partners in the nonprofit sector.

Chapter 7

Lack of Investment in Technology

It is often the case that nonprofits often struggle to secure the funding and expertise they need to upgrade their computers, let alone embrace new tools like AI. According to a Chronicle of Philanthropy 2025 survey, that's in part because funders may discourage or outright prohibit the use of grant dollars to cover general operating costs like software licenses or cybersecurity services. As a result, many charities have been reluctant or slow to integrate technology into their work.

Which leads us to one of the primary barriers to broader adoption and integration of technology: spending, or lack thereof. This is in part due to the financing systems upon which nonprofits rely; namely donations and grants. Funding designated toward technology from the philanthropic sector to support nonprofits is lacking. "Many funders don't fund tech, period," says Alethea Hannemann, CEO of Board. Dev, which connects tech leaders with nonprofit boards. "Those that do really want to fund the innovative stuff, and so some nonprofits are feeling pressured to experiment with AI when their foundational tech" — like cybersecurity or basic software training — "isn't strong yet."

As a result, technology spending in the nonprofit sector represents approximately 2-3% of total spending. Private sector companies typically spend between 3.6% and 5.8% of their revenue on technology. Even as a percentage of overall budget, the nonprofit sector generally spends a smaller percentage on technology compared to the for-profit sector (average of 1-4.4% vs. 3.6-5.8%). This could be partially due to the reasoning mentioned earlier around the reluctance of foundations to provide financing for new software or technology.

For-profit companies are increasingly investing in revenue-generating technologies and digital transformation, while nonprofits tend to focus more on operational efficiency, mission delivery, and donor management systems. Spending on hardware such as desktops, laptops, cell phones have historically been the largest segment, representing about 38% of nonprofit technology spending in 2024. However, software is the fastest-growing segment. There's a notable trend of nonprofits shifting more of their technology budgets from hardware to software subscriptions and cloud-based solutions.

Interestingly, smaller nonprofits spend a higher percentage of their budget on technology, though larger organizations spend more in absolute dollar terms. Nonetheless, the amount of spending is quite low compared to other expenses. Very large nonprofits spend an average of $235,445 on technology annually while large nonprofits spend $101,064 annually. The amount of spending continues to decline, with medium-sized nonprofits spending just $45,184 annually while small nonprofits spend $7,595 annually.

According to the 'Key findings from the 2024 Nonprofit Technology Trends Survey' from Sage Consulting, foundations are investing internally in:
- Cloud services and software subscriptions (the largest growing category)
- Cybersecurity and data protection
- Financial management and grants administration systems
- Communication and collaboration tools
- Data analytics and reporting platforms.

That said, a 2024 survey of foundations from Fluxx showed a reduction of overall spending on technology, when averaged as percentage of budget. The same survey showed that 50% of foundations allocate between 1-5% of their operating budget to technology, while 86% are spending 10% or less on technology. Amounts, according to Fluxx, 'which…raise some flags in the future of technology investment' among foundations.

CHAPTER 8

Lack of Utilization

According to a 2025 Chronicle of Philanthropy survey, 'most nonprofit leaders recognize that up-to-date tech is essential, with a full 64 percent of them saying improved use of technology is among the top three priorities for their organization. The survey found that 77 percent of nonprofits expect to use AI within three to five years, despite only 46 percent using it currently. With the ubiquity of AI tools and our most common software applications embedded into these systems and making it easy to access and use, we can start to incorporate AI into our work.

According to Salesforce's 2022 global survey of more than 1,600 nonprofit managers and executives across seven countries, 74% of nonprofit leaders view the adoption of technology and the ability to leverage the capability of technology as essential for progress. However, the same study revealed a significant capacity gap: only 12% of nonprofits said their organizations were mature in how they use technology to transform their organizations.

Solutions tailored specifically to philanthropy are now available. Most of these solutions focus on the grantmaking process.

CHAPTER 9

Hesitance to Use AI

According to a 2025 Chronicle of Philanthropy survey, 'most nonprofit leaders recognize that up-to-date tech is essential, with a full 64 percent of them saying improved use of technology is among the top three priorities for their organization. The survey found that 77 percent of nonprofits expect to use AI within three to five years, despite only 46 percent using it currently. With the ubiquity of AI tools and our most common software applications embedded into these systems and making it easy to access and use, we can start to incorporate AI into our work.

According to Salesforce's 2022 global survey of more than 1,600 nonprofit managers and executives across seven countries, 74% of nonprofit leaders view the adoption of technology and the ability to leverage the capability of technology as essential for progress. However, the same study revealed a significant capacity gap: only 12% of nonprofits said their organizations were mature in how they use technology to transform their organizations.

Solutions tailored specifically to philanthropy are now available. Most of these solutions focus on the grantmaking process.

In 2025, a Pew survey on the use of AI at work found that 81% of the workers sampled were considered non-AI users. These respondents said little or none of their work is done with AI. 17% of workers sampled said they hadn't even heard of AI being used in their workplace.

In terms of broad acceptance of AI and its capabilities, we have work to do. According to recent study by the Pew Research Center, 'experts are far more positive and enthusiastic about AI than the public....AI experts we surveyed are far more likely than Americans overall to believe AI will have a very or somewhat positive impact on the United States over the next 20 years (56% vs. 17%).' The report continues to say that 'while 47% of experts surveyed say they are more excited than concerned about the increased use of AI in daily life, that share drops to 11% among the public. By contrast, U.S. adults as a whole – whose concerns over AI have grown since 2021 – are more inclined than experts to say they're more concerned than excited (51% vs. 15% among experts).'

It goes without saying that many nonprofits are still early in their digital transformation journey. According to the 2022 Salesforce Nonprofit Trends Report, 'only 12% of nonprofits worldwide are considered "digitally mature," although nearly 74% view digital transformation as essential to their success. Most nonprofits recognize that technology is critical for their work, with 76% of employees acknowledging it as essential to their success.' Nonprofits spend most of our tech dollars on hardware. In 2024, we started to shift toward spending more on software. This spending needs to come with dollars for training.

As younger people do come into the sector, we must have a ready and willing technology platform for them to integrate into otherwise, digital natives will quickly leave and or won't go into the nonprofit sector if it is viewed as mired in old ways, reluctant to change or worse yet using antiquated processes to obtain, share and act on information.

Nonetheless, within our sectors, a debate around the role of AI continues. Some are concerned about privacy, some are concerned about resources necessary for adoption, other voice concerns around bias while others just have no desire to change their ways. Foundation leaders may be concerned that applicants are 'just using AI to write grant proposals.' A foundation telling an applicant they can't use AI to complete their grant application is like a foundation telling an applicant they can't use the web to research their proposal.

Entire movements have been created by the philanthropic sector to alleviate the administrative burdens that come with writing, researching and reporting on a grant. It is hypocritical to blast new technology which de-facto decreases the burden on grant applicants. It's like a university professor in the early nineties telling their students they couldn't use the new word processor capability of a computer but rather had to use a manual typewriter for their papers.

We are starting to see solutions emerging in the sector. Foundations such as the Patrick J. McGovern Foundation are incorporating AI into their operations and offering AI-based solutions to their grant applicants.

Even so, nonprofit leaders may be uncertain about how exactly they can leverage this emerging technology, but they indicate they are ready to embrace the revolution.

CHAPTER 10

Isolated Databases

The lack of investment in technology and a slow rate of on-boarding technology leads us to the next barrier: the inability of our current systems to communicate and synthesize data. Dan Pollatta's 'The Flat Org Chart' singles out the lack of 'data-gathering, incidence mapping coordination, data analysis and external communications' as one of the single greatest limiting factors to progress in the nonprofit and philanthropic sectors.

Currently, most of our data exists on our own servers and within our own subscription-based software programs such as Donor Perfect. We can individually access this data, but it can't be accessed by anyone else other than the company that provides the software. Our grants management systems are located internally to our own foundation and do not 'talk' or share data with other systems. When a county or state wants to determine the total number of dollars donated to a certain cause, there may or may not be a website with this number. If there is a website, the number is probably at least a year old.

Funders often require the use of certain databases for client records or case management purposes. While well-meaning and intended to shift away from paper and foster stronger data management capability, the result is several fragmented systems that don't communicate with each other and are not connected. When we try to capture data on the impact of our work, users in the nonprofit and philanthropic sector are left to cobble data sets from one system and reformat or reconfigure entire datasets for use in another system, instead of being able to cull insights from multiple connected data systems.

Rarely, if ever, is this data shared among foundations in a way that can be digitally scaled and globally aggregated to demonstrate the impact of our sector. It's hard to determine the impact of our work collectively if we don't aggregate and share our impact globally. A more wholistic set of data points that are universally recognized and shared would allow for this type of performance aggregation or performance analysis. Compare this to the financial sector where investors, corporations or financial advisors don't need to wade through an IRS site to find performance data on companies.

Granted, there is no oversight agency which looms over foundations, threatening to take their entire corpus if they don't meet their stated mission. The only real oversight from the IRS comes in the form of spending policy which means that in order to maintain their charitable status they must grant out a certain percentage of their corpus annually.

To be fair, some foundations, in the interest of transparency, share their reports online in order to inform the field and help others who have similar missions. Foundation staff frequently share both empirical and anecdotal information about agencies they fund with other foundations. But that's the extent. There is no clearinghouse or electronically connected data warehouse that captures the collective performance of nonprofits. Later in this book, we will see that this is one of the greatest barriers to reimagining our sector. Nevertheless, it's a barrier that has been overcome in other industries such as education, health care, housing and environmental protection.

Many of the existing donor and grants management databases are essentially souped-up Access databases. Some of them are now adding chat bots to make the experience seem more advanced. This functionality makes it easier to use the software or database, but it does nothing to transform data and the way we use data to make decisions. Furthermore, rarely do these datasets or software systems communicate with other systems and therefore insights other than what is available within the client's sphere of information are limited.

Add into the mix ill-informed and outdated expectations around individual donor and client data privacy and protection and you have isolated data systems. Creating organizational or systems-level momentum to change these systems is daunting if not impossible for an individual or organization. Such technological transitions inevitably come with a cost and can oftentimes suffer from a lack of ownership.

Most of this information comes directly from each institution. All of us use databases but most are monolithic or singular in that they do not connect with any other networks or databases. The data remains diffused and is not cross tabulated to aggregate a broader impact schematic that would be useful to donors and other actors in the sector.

CHAPTER 11

Centralized Financial Data

The centralization of our financial data created by the IRS is an impediment to progress. All tax-exempt nonprofits must make their three most recent Form 990s available publicly. Form 1023 which are applications for tax-exempt status also must be shared publicly. This information is usually submitted at the end of the fiscal year and is therefore backward-looking. By the time the IRS receives the information, and posts the document on their website, a delay of a year to 18 months can pass. Thus, IRS data is barely usable for anyone other than an IRS compliance officer or an accountant.

Currently, key information on nonprofits and foundations is held in a few sites controlled by either the IRS or corporate entities. Among nonprofits and foundations, no centralized reporting authority exists other than states and the IRS. States often require charitable organizations to register while all corporate entities are required to file updated 990's to the IRS. It is possible to access this information, but often only up to certain level of information. In the case of the state registration, this data can be limited to the name and address of the entity.

Nonprofits file their IRS 990 annually showing where they received money. The IRS makes 990's available through their website. Users can access what is essentially a screenshot of a PDF file. The forms required by the IRS only ask for financial information, salary of three highest paid employees, names of board of directors and mission. Nonprofits must provide a list of the funders and their grants received in the past year.

Other online platforms such as Candid draw from this information and make it available for free, but with the requirement of an email and login to access the information. Candid also provides the amount of grantmaking for foundations, but this information is only available in a PDF format which is a cumbersome file format for sharing information. Therefore, these sites are marginally helpful but in the end are mostly just depositories of data showing where and to whom someone gave along with the recipient of the funds.

There is no centralized database which weaves the information we submit to the IRS together. Nonprofits and foundations are not even required to provide information in a format which can be digitally scaled.

CHAPTER 12

An Opaque Culture

Foundations are not required to share board meeting minutes, nor do we need to explain why we are funding an organization. But within this broader debate on reimagining our work using technology, let's pause and consider the question: *Is the level of privacy foundations enjoy a feature, or a bug?*

As Dr. Rob Reich from Stanford states in his book, 'Just Giving: Why Philanthropy is Failing Democracy and How It Can Do Better.' Foundations remain 'black boxes, stewarding and distributing private assets for public purposes, as identified and defined by the donor, about which the public knows very little and can find out very little.' We will hear more from Dr. Reich as we progress through the book and how this lack of accountability is one of our greatest enemies.

There is an underlying air of hypocrisy, or for a lighter touch, *inconsistencies in the approach*, when it comes to transparency among foundations. Clara Miller pointed out, in 2022 that 'foundations champion transparency on the part of grantees as well, insisting on measures for effectiveness, program performance, financial management, and evidence that grant dollars are used efficiently.' She went on to point out how foundations 'strictly control data and narratives about their own performance in areas such as investment, staff, strategy, and allocation of funds. They generally do little more than the bare minimum required by the IRS and state charity regulators, which means making their latest tax returns, and sometimes their audited financial statements, available.'

While this opaqueness may serve to protect donor privacy and granted is perhaps just the result of a lack of requirements around transparency, what has it really allowed the sector to accomplish? How many patents or intellectual property examples can we point to within the sector?

PART 4

The Overhead Myth

Exasperating the slow nature of change inured in the sectors are more cultural; philanthropists and donors generally retain a misconception that nonprofits must be frugal in investing. They should be lean or operated on a shoestring. There are multiple reasons for this line of thinking, mainly the charitable mindset that plagues our approach to financing within the sector, and the broader assumptions underlying 'charitable' organizations.

As a result, the number of agencies with the capacity to adopt ready-to-scale solutions to the greater societal and global challenges discussed earlier is limited. Pair this with the enormous amount of charitable dollars earmarked for philanthropy and you have a mismatch between donor expectations and the ability of the nonprofit sector.

As illustrated in Sean Davis' 'Solving the Giving Pledge Bottleneck,' even though a nonprofit may seek to solve a problem much greater in size than their own agency could possibly tackle, the deficit mindset inured in the sector leaves nonprofits lacking for planning and growth management skills. Thus, many well-meaning nonprofits lack a reasonable and ambitious growth plan.

As Davis points out, 'philanthropists looking to give billions… are having a difficult time giving it away.' This is an issue of scale. While the amount of money donated by individuals, foundations and corporations is growing, it is 'diffused across 1.5 million nonprofits.' As nonprofits are accustomed to doing more with less while placating the donor who wants overhead expenditures held to an unreasonably low-level, 'not that many nonprofits can readily take in and quickly expand their operations' with large donations.

Misconceptions around overhead abound. 'Nonprofits shouldn't spend more than a certain amount on overhead as donors what their donations to go to programs that provide impact,' says the well-meaning foundation board member. Therefore, nonprofits take great pains to restrict the amount of overhead that appears on their financial statements. Capital investments and investments in staff are seen as excessive. Market-rate salaries are considered too generous.

This results in the never-ending cycle to fund programs and not infrastructure. Our capacity to experiment with new and unproven technology is limited. With a lack of capital free for research and development, forget building anything new, other than a building.

Layer onto this the generally accepted, broadly adopted but to some degree eroding belief that there is a magic 'overhead' metric that organizations should not exceed in order to be 'efficient. This cycle is further perpetrated in part by overly simplified rating systems and the like (more on that later.)

Putnam-Walkerly also makes the key point that philanthropists or donors can't spend too much time discussing innovation if they themselves or their foundations are not innovating as well. If you can't innovate yourself, how can you recognize it in others?

Donors demand immediate impact and focus on confining metrics like 'program costs' while leaving out any room for nonprofits to take risks. There is a lack of incentives for nonprofits to take risks. Research and development investments are hard to justify. Nonprofit ratings sites don't measure a nonprofit's impact so there is another disincentive to report to donors on true impact.

Sites abound that cater to the inquisitive nonprofit donor. However, calculating the overhead of a private foundation is an entirely different exercise. Foundations can mesh program and grantmaking costs with overhead and payout. The process for calculating this number varies by foundation. Comparing the overhead costs from one entity to another is like comparing apples to oranges.

CHAPTER 13

Slow Pace of Change

Nonprofits, too, become inured to change and spend time trying to please the donor's ever-wavering priorities and cause du jour. How many businesses change their model every 6 or 12 months? How many businesses develop a new line of service without dedicated resources overnight, as nonprofits sometimes find themselves called to do by a funder? In some ways, pivots in the private sector can be easier, but not so in the nonprofit or philanthropic sectors. Pivoting on a dime in an under-resourced environment where donors and funders must be pleased can be difficult.

Our earlier point from Putnam-Walkerly that philanthropists and donors shouldn't spend too much time discussing the need for innovation if they themselves are not innovating is a reoccurring theme in philanthropy. A lack of meaningful spending and support for spending and adoption of technology is a missed opportunity which can no longer be overlooked. During the pandemic, many nonprofits and foundations were forced to shift much of their operations online. Now that we have transitioned and increased our awareness around technology, we are unlikely to shift back in a wholesale manner due to lack of funding. Doing so would be both a missed opportunity and negligent.

CHAPTER 14

Due Diligence Grind

Researching and identifying a good cause is called 'donor diligence' and is part of the due diligence process Due diligence takes time and effort. If you would like to donate to a nonprofit, you go online search by keyword and perhaps location to identify a few local agencies. You might even volunteer to get a better understanding of their work or perhaps the cause is close to your heart. Perhaps a few friends have a favorite nonprofit or someone at your church supports a local group. After some thought and research, you donate online or via check. Or you hear a news story about the inferior performance at a local elementary school. You go online to research who might be working to improve the conditions as this school. You find an agency that works directly with this school, and you donate $100 to their cause. If you are an especially knowledgeable donor, you can tap into the online clearinghouses designed to rank and provide data on nonprofits. With the myriad number of nonprofits and causes, it can be time consuming to figure out what and who to support.

CHAPTER 15

Grant Applications

If you are a nonprofit seeking funding, you maintain a catalog of grants and seasoned Development Officers (fundraisers) who cultivate donations from individuals and foundations. There may be a grant writer on staff who researches potential funders and writes grant applications. They will submit grant requests to foundations whom the grant writer has identified as having synergistic interest. According to the Center for Effective Philanthropy, the average grant application takes 27.5 hours to compile and submit.

Sometimes sending out a grant application feels like putting a message in a bottle. Who knows when or if you will get a response, nor do you really have any idea about where the bottle is going or what forces will carry it along. Foundations can take weeks if not months to make funding decisions. Not without merit, as these are important decisions that usually are balanced against other competing and equally as important proposals.

It takes a nonprofit time to complete grant applications and report back to the funder. Add in events to recognize existing and cultivate new donors to these efforts. Finally, there is the reporting. Reports go to donors. Reports are required on regular intervals to foundations. Reports also go to the IRS annually. There is an annual audit. It's a given constant that a report due to someone. All in the name of quantifying performance and letting the funder know that their funds are being put to good use.

If the nonprofit is well managed or if they are charged with using a designated database that the funder provides (usually an old clunky Access based system that takes inputs and nothing else), they can pull data from these systems or spreadsheets they capture through the course of their work. If not, there is a race to gather data from different individuals and departments to meet a reporting deadline. In the end, most of the time the data is put together to satisfy the requirements of a single funder.

Conversely, working at a foundation can involve reading and analyzing grant proposals along with endless prep for another round of approval of grant applications. Add the research on best practices and time spent staying on top of developments in the sector and the days quickly pass by. Inevitably this involves using the information provided by the grantee via the grant application, researching other sites to verify the efficacy of the proposed program model or proposed service and combing through the grantee's website. Conducting due diligence and making funding decisions in this manner is a barrier that can easily be overcome with an efficient application of technology.

CHAPTER 16

Reliance on the Charitable Tax Deduction

The charitable tax code focuses on donor finances and mechanisms for giving and does not consider whether a donor gives to build a new library at a university, house 200 homeless people, or create a new park. Regardless of where it goes, the charitable 'write-off' for any amount remains the same. Whether the gift is impactful or truly relevant or necessary to alleviate suffering plays no part in our charitable tax code. Changes in the tax code can take decades and usually amount to incremental tinkering around the edges.

Taxes focus on the actual gift, not what is done with the money afterwards. Rather than hire staff or cut through the data online, it's often easier for a very wealthy individual to just donate a substantial sum to their own foundation or a Donor Advised Fund (DAF) held at a foundation.

DAFs are vehicles established to take money from donors and hold the proceeds in another foundation. In theory, the money can be distributed as needed or as a cause arises. DAFs tend to create another layer of bureaucracy to the process of giving and accessing information. Financial Institutions or foundations which manage DAFs generally advertise the amount of funds they hold and discuss their impact as a secondary feature. As outsiders, it can be difficult to determine the impact of the dollars they give out.

Because we don't gauge the level of charitable deduction on the degree of impact provided by the gift, there is no performance motive for the system. Further, donors have no real motive to ask how their gift affected an individual or a program, so long as they receive their tax write off.

Furthermore, recent Congressional hearings in the Summer of 2025 focused on nonprofits should be a warning to the sector that it may be time for a re-examination of the charitable tax deduction. The tax deduction is not the primary reason people give. In fact, according to the Philanthropy Roundtable, while 2/3 of US households donate to nonprofits annually, only 1 in 13 claim the deduction on their taxes. Developing increased accountability and transparency around how nonprofits use money and how the money held in DAFs is distributed should be a priority for the sector.

CHAPTER 17

Lack of Financial Transparency

Which leads us to our next barrier to reimagining giving, which is our reliance on the traditional financial system. Foundations make grants and send a check to a nonprofit. Nonprofits use the money to purchase labor to provide programs. This flow of capital is difficult to follow because we have no real-time mechanism to track donations.

Our reliance on the traditional financial system means that it can be difficult or impossible to know when money goes into the system and what happens to the money once it is received. We don't know at any point in time where capital is flowing, who is benefiting and how people are benefiting from charitable dollars .

Even for the individual donor, once you donate to a nonprofit, unless you conduct a forensic audit, at the granular level it can be difficult to discern exactly where your money was spent and who exactly benefitted. If you donate to a feeding program, there is no way to tell what individual was helped with your donation. This is what drives the popularity of donation platforms such as GoFundMe, connecting directly with the beneficiary.

CHAPTER 18

Reliance on the Traditional Financial System

Ready access to banking and financial services is something many of us take for granted and never question. Access to credit and loans is essential to economic mobility and opportunity. But not everyone lives near a bank nor can everyone even access credit from a bank. Black and Hispanic Americans are more than twice as likely as white Americans to be unbanked or underbanked. There are a number of reasons why, discriminatory practices and regulations being one. But there are other reasons as well. If you are on the run from someone like an abusive spouse and living on the margins of society, or perhaps you lost your identification or maybe don't have an address, it can be hard if not impossible to get access to the banking system.

In the US, the banking system has been criticized for being inequitable for years. Red-lining practices which prohibited black people from taking out mortgages in white neighborhoods were common practice and led to neighborhood segregation that still exists today.

Without access to the traditional banking system, people rely on high-priced money transfer services, expensive check-cashing stores and payday loan services. These businesses can charge up to 10% just to cash a check or send money. It's easy to end up in a spiral of debt once you use one of these services.

Requirements for identification, especially today with increased scrutiny, can be another barrier or just something that makes people think twice about using the system.

PART 5

Reimagining the Donor Experience

"It is more difficult to give money away intelligently than to earn it in the first place."
ANDREW CARNEGIE

To reimagine philanthropy, let's start by reimagining the way we give. Let's look specifically at some of the changes we can make which can benefit not just foundations, but individual donors and the nonprofits we support.

In 2018, when I first started writing pieces and blog posts which would morph into this book, one of the things I was thinking about quite often was access to data. This was mainly given my position at the time working for a mid-size private foundation. One of our key roles would be to access data on a nonprofit applicant as part of our due diligence.

This then led me to think about the donor who provides large sums (or even small donations) to a nonprofit. What kind of data do they receive? When do they receive it? How do they receive it (via newsletter, e-blast, site visit, etc.)? How do they know the impact of their donation? Are they told by an employee of the organization, do they peruse a website and read about 'stories of impact?'

Visit any large foundation and peruse their website and you may find reports and publications showing the results of their work. Note the use of the conditional 'may' because it is not required for foundations to publish the results of their work. The bulk of words written around philanthropic investing goes toward describing how much and to what cause someone invested. It's not very often that results of these investments are posted.

Individual donors can't see who is receiving the largest donations by with much degree of specificity. Our existing data systems and platforms do not coalesce and connect in a way to offer this information. Most of the information provided is disjointed and usually quite dated. These electronic silos or disconnected pathways and rails that restrict access to foundation spending and ultimately nonprofit performance hinder progress. Greater transparency around the flow of data and money and the impact of our grantmaking will benefit the sector.

Leveraging available technology to penetrate this information and provide it publicly and to our partners across the sector would transform the way we approach donations and grantmaking. This system should be developed at scale and communicated in a way that is easily digestible and instantly understood. Combine data from foundations to show who is donating and to whom. Providing this data on a regular interval, say weekly or monthly, would allow us to also judge when grant dollars are flowing. Finally, make all of this data visible to everyone, in real time. Allowing people to view this data and see the real time impact of an organization will help direct funds to agencies and solutions that are truly having an impact. Weaving this data together and generating collective insights into our work as sector would allow us to harness the collective knowledge of the field.

You would need access to a large volume of data across several organizations. Large nonprofits and foundations already have performance data. Even mid-size nonprofits retain performance data. With market forces and financial incentives (more donations), most nonprofits would become part of this system. Eventually, this practice of sharing data in a transparent way will include all nonprofits and foundations.

Given the right access and inter-connected systems, AI could easily mesh key data points from the IRS, Candid.org, social media, chat boards and ratings sites to form a meaningful decision point to assist program officers in making funding decisions. But to truly harness this power at scale and bring about a true digital transformation, we need more connected datasets and more transparent, readily available and accessible information about nonprofit organizations. If this data is siloed by organizations, as it is now, it will continue to be both difficult and expensive to obtain insights on the impact of your work.

CHAPTER 19

The Flow of Donor Dollars

Like a train moving along a train track in a pre-determined direction, many investment professionals refer to the infrastructure financial systems rely on to move money as 'the rails' of the system. Finance professionals and even everyday investors can see these rails, thus the system is built on transparency and accountability. Granted, this is a demand of the investment world, transparency and accountability.

Data also moves on rails that facilitates the transfer, storage, and retrieval of data; ensuring that information moves efficiently from one point to another. Data rails or data pathways are crucial for maintaining data integrity, security, and accessibility in various applications and systems.

The programmatic and financial rails in the philanthropic sector consist of one rail going in one direction: from the foundation directly to the IRS in the form of our required annual report and Form 990. Individuals, nonprofits and foundation program officers can't see or access the information flowing on these rails. Individual donations also flow in one direction: from the donor to the nonprofit. The rails are not public, nor do they update in real-time. As a result, we can't see where or when grants are made, who is receiving grants or who is making grants, until long after the reports are submitted to the IRS. That is unless the grants are reported on the foundation's website.

With the advent of the online 990 and searchable documents we are getting closer, creating transparent rails within the nonprofit and philanthropic sectors. Nonetheless, it's still incumbent upon the individual to do their own research and ask around.

A critical deficiency within the philanthropic sector is the information black hole around where grant dollars and donations go, to what cause, where and to whom, *in real time.* In the next few years, this will likely become one of the major issues holding back the field. (When I started writing this piece in 2019, I would not have written that last sentence. But now in 2025, we are starting to see the repercussions around the lack of transparency within our field.)

CHAPTER 20

Donor Accountability

So why don't we use the large data sets which are already captured and held by the IRS and large corporations to build a system designed to help people make donations, large or small? Why don't we start using this data to help people make an informed decision about when is the right time to donate? We could create donor profiles that both verify whether donations are made and which can also tell us who received the donation.

How would we obtain the data? Obviously to do this we would need access to data in real time. This can be accomplished by integrating systems and other data platforms which provide information such as the amount of money from others going to a cause, the impact those funds are having and the agency's digital footprint (how many people are talking about the agency online or referencing their work in online discussions.) These datasets are already out there and can be integrated to form a more complete picture.

Of course, this runs into all kinds of arguments around privacy and individual rights. No one can tell a donor to whom or where to make their donation. Some choose to make donations public, others prefer to stay under the radar. According to an article by Mike Scutari, <u>What If Five Top Tech Billionaires Emulated MacKenzie Scott's Approach to Giving?</u> states, 'it's the family's right to keep those figures under wraps, but it perpetuates the <u>cynicism-inducing perception</u> that funders use philanthropy to burnish their brand without providing transparency into their recipients.' The Soros Foundation is probably one of the ultimate examples to this type of 'murkiness must mean malintent' assumption. As we discussed earlier in this book, this perception is starting to catch up with the sector.

We're probably a bit far from that point but systems are moving toward real-time insights on the nonprofit sector. Nonetheless, our existing benchmarks and data repositories are slow to evolve. Only within the past few years did IRS and Candid begin the transition to posting donor data via a 990 onto their website in a searchable format.

Even with the existing information available on the IRS website, if downloaded at scale via an AI powered aggregator such as Google's NotebookLM, we can create a searchable database of all donor data. We could then create an algorithm that regularly accesses the data and posts information online. This data set would come from payment providers such as PayPal or reported weekly or monthly by foundations. Using this information, we can start to see how simple indicators such as the increase or decrease in funds towards certain causes would open a whole new realm of insight. This information could even serve as a leading indicator that a problem or local issue has mitigated or in some instances is no longer a pressing issue that requires charitable dollars.

CHAPTER 21

Seamless Giving for Donors and Foundations

How many times have you picked up your phone to do something, were distracted by the news or social media, and completely forgot why you picked up your phone in the first place? How many times have you tried to make an online purchase, only to forget a password for your account or spent time updating your expired payment info. This is called 'friction', and it has a cost. A cost in lost time, diverted attention and loss of focus which eventually drives the donor away. Friction costs are a common consideration in business transactions. Basically, 'friction' is a term used to describe the machinations or steps someone must go through in order to complete a purchase. Businesses of all types think a lot about how to reduce friction costs so they can make the sale and receive your money as quickly and easily as possible. The last thing a business wants is to put up barriers that keep you from making a purchase. Amazon has mastered this with their 'one click' checkout.

Donating online can be another source of friction. When was the last time you donated online? How easy was the process? How much 'friction' was there? Eliminating frictions in the donation process would go a long way to not only increase donations but create a more knowledgeable and engaged donor.

Researching the nonprofit takes time and effort. Data on the nonprofit is likely spread across multiple websites, therefore you must visit a few sites to verify the group. This research process is called due diligence.

Let's consider the problem of homelessness. If you were inclined to make an online donation to support a group that works to alleviate homelessness, here is a breakdown of the due diligence steps we require of donors if they want to support a local shelter or help the homeless in their area:

1. Google 'homelessness'
2. Seach results for agency in your area providing services
3. Visit agency website (you can start here if a friend tells you about their favorite group)
4. Navigate sector-specific language likely used by nonprofit
5. Verify agency and identify services they provide
6. Visit other websites to check Form 990
7. Try to login into Charity Navigator
8. 8. Find password that you forgot for Charity Navigator
9. Spend time going down a rabbit hole on Charity Navigator
10. Ask why and how Charity Navigator became the authority of good charities
11. Go back to agency website
12. Find 'donate now' button on website
13. Click 'donate now' button to go to PayPal or other payment platform
14. Login and link credit card with outside platform
15. Find your password for PayPal
16. Update your payment info which expired two years ago
17. Input amount of donation
18. Add the required 5 or 8% fee for payment platform provider

Click 'send' and your money now goes to the payment platform and eventually makes it's way to the nonprofit agency in the form of a bulk payment made monthly by said payment platform. How long it takes to get there or how long it takes to manifest itself in beneficial services that help the homeless in your area is unknown and varies greatly by the agency. Lastly, receive info on donation in form of 'thank you' letter highlighting recent work

A more seamless donation process using the generative capabilities of AI would weave together multiple data sources into the decision-making process. These data sources would be integrated from multiple platforms into one seamless point for the donor to access. It would not create friction for the donor and alleviate the laborious due diligence process.

We can see how much easier this process becomes when we look at the steps required using our notification system with a pre-determined donation amount:

Identify causes of interest (Accomplished through your phone or embedded sensors in your glasses which follow your online interest, the rush of your heart rate or eye movement and whether or not you react to your environment and things around you)

Receive notifications (You drove or traveled through an area where homeless people congregate.)

Receive predictive and forecasted impact model of agency(s) working on said cause. (These are push notifications that arrive in your inbox or on an app. Filters for these notifications could be managed up front by you.)

Donate electronically (Seamlessly as most systems already have your credit card info and don't require a login.)

Monitor impact of donation in real-time (You get more than an email 'thank you' but continual notification or access to an interface showing how the agency you supported continues to make a difference.)

Receive on-going notifications/impact indicators/impact reports from agency (This information flows continually to you after the donation is made.)

Notice the process now requires fewer steps and alleviates many of the due diligence steps currently required.

Layers of verification could be built into the system, such as a backup program connected to the IRS database that verifies the agency's nonprofit status. Further layers could verify that they do in fact work in the specific area and that they have been successfully providing services. This verification would rely on transparency from the agency and the allowance of full and total access to records of where and how they work. All this verification is currently the responsibility of the donor but would be transferred to the agency and ultimately verified using existing data already available. This transformative model could allow us to give and identify causes to support in a consistent manner and could easily combine all these steps via an app on your phone. The model facilitates electronic donations through seamless payment mechanisms that connect directly with a nonprofit, bypassing existing payment processors or bank servers.

Obviously, this is an economics issue to some degree. Some agencies operating on donations may not be able to afford to send location-based ads to mobile phone users nor may it be ethical to identify known areas where the homeless tend to congregate with a digital footprint. Nonetheless it could be a higher-level touchpoint for nonprofits. Instead of the bulk mailing to thousands or email blasts to website visitors, perhaps it's a more real-time targeted method to inspire a donor while they are looking to be inspired. You can take a few minutes to make a donation while you are moving along or once you reach your destination.

We could use this same seamless mechanism to support environmental causes. For example, a donor may be driving by a toxic waste site. The app could initiate a similar process as above.

By providing real-time information, we can enhance donor interest and ultimately increase donations while lowering barriers to giving. This donation and information can be integrated into a dashboard, utilizing algorithms to display the amount of money allocated to a cause and the impact of those funds.

When researching a cause, instead of banking on the ability of a nonprofit to deliver, we can use the data on donations via the same platform to develop predictive impact models allowing for informed decision-making to guide their donation. For larger donations or in the case of a grant, the models can scale up or down depending on the size of the donation, allowing the donor to determine the amount of impact based on the size of their donation.

Real-time monitoring of the impact of these donations is also a crucial feature. Finally, after the chosen cause is supported, users receive ongoing notifications, impact data, and comprehensive reports related to their selected causes, fostering a continuous connection and engagement with the issues they care about.

Our current software systems are a limiting factor in our work. We need a wholesale transformation in the way we capture, utilize, communicate, display, monopolize, manage, and present data. We need to move beyond the computer and screen into a real-time domain built on experiences.

CHAPTER 22

Cause Du Jour and Donor Signaling

According to Wilfred Chan the Editor of New Public Magazine, money is power and 'the way power is distributed, accumulated, transformed, and dispersed at every layer of society requires more transparency.'

There currently exists a social graph of everyone around the world who is on the internet. This is your 'digital profile.' Guess who owns your digital profile? Yes, the corporations that own the sites you visit. Guess what they do with your profile? They sell it to advertisers.

It's important to share the results of our grantmaking. If we don't share more with our partners, we continue to hold back the field. Otherwise, we are held hostage what the concept that Lucy Bernholz's introduced in her work on creating and transforming philanthropic markets; 'endorsement philanthropy.' Which basically creates an environment within the philanthropic sector whereby large foundations tend to have the most knowledge, but the information is not readily shared or made available in a meaningful and insightful way.

After the murder of George Floyd there were countless new articles covering the amount of money corporations pledged toward the Black Lives Matter movement. Tracking who pledges what is the easy part. Initially these pledges of support were all great press but after a few years passed and you took the time to peel back the facts from the headlines, much of the funds were diverted not to organizations on the frontlines but to more traditional causes, such as lending for new home owners. A <u>Washington Post analysis</u> of racial justice pledges after George Floyd's death reveals that only $4.2 billion of the $50 billion in corporate donations went to organizations specifically committed to criminal justice reform.

To follow through with thoughtful due diligence and cut a check is another matter. Rarely do you see a tweet from a major corporation saying, 'just finally cut that $25 million check to Black Lives Matter today.'

Public proclamations with no accountability of proof that someone actually followed through on their pledge can be misleading. Someone may think 'the (insert name of corporation) is giving to this cause, so it must be worthwhile, and it must be a good use of funds, so why shouldn't I do the same thing?' Who wouldn't fault the donor for following the lead of a smart, international company with teams of analysts pursuing data and the latest research to identify the best grant recipient?

The only way to truly verify whether an individual or a corporation donated is by reading about the donation in the news. This lack of verification can lead to confusion and obfuscation on behalf of the donor.

CHAPTER 23

Connecting Donors with Causes

Nonprofits rely on financial support from many sources, including philanthropy, government and individuals. Maintaining positive, meaningful relationships between donors and nonprofits are critically important.

Using technology, we can transform the donor experience into something transcendental and much more meaningful. Bryan Stevenson said that by "get [ting] closer to people who are suffering... you will find the power to change the world." He came to this realization firsthand when he took on the prison system in college and had to spend time with inmates. He noted that the context of a course, he spent a month working with death-row inmates. "I got proximate" to the prisoners, he said. "In proximity to the condemned, everything changed for me. I found a calling."

The poverty experience is a day-long immersion training about poverty. The goal of the training is to teach someone what it is like to live in poverty. Living in poverty isn't just about not having enough money, it's about the daily challenges that arise as result of living paycheck to paycheck. In the poverty experience, you are grouped with a cohort of others in a room. A separate cohort is charged with facilitating the experience. You are then given scenarios that you have to play out in real time throughout the day. For example, you have $5 in your pocket, and you have to find a ride to the DMV to update your license.

Much of the time is spent walking around a room with other participants. There is a lot of verbal interaction with other actors or hosts. One host may represent a payday lender. Another may represent a bus driver or an employee at the DMV. In order to navigate through your day based on the situation you are provided, you may or may not have to work with one of the hosts to solve a problem specific to your situation. There are all kinds of bureaucratic, logistical and administrative barriers thrown up as you try to navigate through your situation. But in the end, you know it is just make believe.

Now imagine conducting this same experiment but using VR. You could get a much more realistic experience if their entire sensory experiences were projected in VR. You could literally be out on the side of the road, walking to a bus stop while traffic whirs by you at 50 mph. We could connect neuro sensors to provide an image of and feel for the environment, hot and windy. Dry and loud.

This virtual experience helps people to come closer to the experience of the person they are trying to help, to 'get proximate.' Becoming 'more proximate' is a key element of grantmaking. Oftentimes foundation staff and trustees to be a degree or two removed from the lived experience of the people they are trying to assist. A virtual experience such as this will bring them closer to the people.

CHAPTER 24

Case Study

A Seamless Process to Help Alleviate Homelessness

In 2023, I was riding the Brightline passenger train from West Palm Beach to Miami. The train goes along the east coast of Florida and skirts the edge of towns, mostly old coastal areas of larger built-up metropolises like Fort Lauderdale. As such it goes through some rough stretches and industrial areas.

At one stop along with way I looked out the window and saw a gathering of old mattresses. There was debris and detritus strewn around along with some old items of clothing. At first it looked like someone had used the area for a dumping ground or perhaps a dumpster was overturned. This spot wasn't one that was tended to very often by the city and clearly no one had been or would be coming along here to pick up the debris.

Looking more closely at one of the mattresses, I noticed a human being wrapped up in a blanket. I couldn't see his or her head or any other parts of their body but clearly there was a living breathing human being under the blanket sleeping on a mattress out in the open at around 10am in the morning during the week. This person wasn't out there for pleasure or for the experience. They were out there because they had no choice. This was where they chose to lay down the evening prior to rest. Maybe they were strung out. Maybe they were sober and just chose this spot strategically knowing that no one would come by to bother them. The only people who would see them here were sitting next to me on this train, passing through to other areas.

Although I only saw the person for a brief moment I wondered if there was a way to help them. I couldn't nor didn't want to disembark the train. I didn't want to go and wake someone up while they were sleeping. This was their area after all even though it wasn't private property they had clearly staked out their claim to this spot of earth debris and trash strewn as it was.

We've probably all come across a homeless person. In some ways it can be jarring enough that you divert your gaze and act like you don't see them. So many questions cross the mind when you come across a situation like this. How can you help? Is there more you can do for someone like this besides handing them money and wishing them well? Who else can help them? It's hard to know what agency will or can help them. Which one operates in that area? Which one offers help without judgment or required commitment?

We come across situations and problems like this all the time. Sometimes we may be so inclined to go home and research the issue, find a nonprofit in that area that has some verbiage on their website about helping the homeless or helping solve the issue you're looking at. You peruse the website and if it looks legit you may offer $20 or $50. Shut off the computer and go about your day. Not that this is a bad thing or is anything that should bring about guilt. You want to help. You're not asking for anything in return. But how do you know your money actually helps that person? How do you know when the agency uses that money for that specific purpose? How do you know if anyone else is also supporting this specific cause? Unless the agency tells you directly, you don't.

Marketers use a term called 'geofencing' to describe the process of sending emails or ads to individuals when they are in proximity of their business. By connecting and aggregating data from wearables, sensors and AI onto a central platform, website or app, and combining this information with the geofencing capability provided by our phones, we can envision a more seamless, almost natural, donation process. We know where many homeless individuals are and where they live or congregate. Most communities conduct an annual 'homeless count.' Volunteers go out into the community to known areas where the homeless congregate and count the number of individuals.

Using existing datasets, we could design an app and use a geofencing method to send information to individuals when they are near an area where the homeless congregate. The app would not identify the exact location of a homeless individual but merely serve as an educational tool to make donors realize that the homeless are more prevalent than they may realize and that they probably pass by homeless people in the normal course of their daily routine more than they realize.

If you are inclined to support the homeless in your local area, you could download the app and agree to receive push notifications. The app would also match you with a verified and trusted agency in the area working to alleviate the suffering of the homeless. You could even agree to donate a set amount instantaneously to these very agencies by agreeing in advance and setting a dollar limit for your donation in the app. The issue in a particular area would be solved when you can drive by without receiving notification.

There wouldn't be any personal data or information shared via the app. Understandably, homeless individuals prefer to stay out of sight, congregating behind buildings or in other areas. The app would only know when the user (donor, program officer or otherwise) is near an area where the homeless congregate. Integrating more advanced technology such as smart glasses into the process could only enhance the amount of information we receive.

Instead of the individual having to go and research the issue and find the agency helping to solve the problem, the data flows to the individual organically. We can then choose to support the cause instantly, instead of hours or days later when we finally sit down in front of our computer.

CHAPTER 25

Visualizing Donations

Another way to leverage our collective knowledge and experience is to create a new field that models the financial sector. Within the financial sector, to drive more money and increase their own profits, financial advisors constantly endorse their chosen company or cause. You can flip on CNBC and hear someone endorsing a company or sector.

In the financial markets, the flow of capital is a commonly used indicator to determine levels of investment in general, within certain sectors or specific companies. In general, more money going in means increased investor confidence. The inverse is also true; more money flowing out of a specific sector or a company could point to distrust or a perceived lack of performance. By monitoring this flow, we can instantly determine the financial health of a publicly listed corporation.

One of the most visible indicators of financial health is the stock price. There is generally a story behind the movement of a stock. Right or wrong, we assume that if their stock is going up, the company must be doing well or is perhaps in the midst of instituting a change that will lead to stronger performance. The inverse is also true, if a stock is dropping it may indicate that earnings are going down or perhaps the successful CEO just announced her departure.

Maybe the earnings were just released, and they beat market expectations. Maybe the CEO just announced she is unexpectedly 'leaving' next week. With no replacement. The ability of the financial sector to share this level of detail is largely due to the fact that information on company performance and where investments are going is shared in real time.

While we don't necessarily require the same level of insight when considering whether or not to invest or make a donation to a nonprofit, why can't we use this same model with nonprofits? Why can't we have a space where monetary or philanthropic endorsements are publicly shared in real time? Instead of the S&P 500 index showing the performance of companies, what about an 'impact index' that we can track in real time to show the performance of charities and nonprofits?

As in the financial markets, donors also tend to follow headlines. They read about an interesting program or a nonprofit doing good work. Perhaps their wealthy friend brags about their favorite charity at a cocktail party. They then write a check or make a grant to support the work.

But in the philanthropic sector, we only know if a philanthropist endorsed a nonprofit is through first-hand knowledge, by an announcement on a website or if we read about online. There is no universal framework offering a real time exchange of information that can be structured in a way to signal an endorsement. As we discussed earlier, we can't see where the donor allocations (investments) are going until they are announced publicly by the nonprofit or in some instances by the donor.

Just as CNBC has their 'ticker' that scrolls along the bottom of their screen, the nonprofit and charitable sector would benefit from a real-time indicator showing where donor and grant funds are going and how it is being used by nonprofits. Access to this information would represent a true transformation of the sector. Providing ready access through instant reporting would preserve the integrity of our sector. Already used at the highest levels to obscure support.

In order to achieve this level of information and transparency, we first must acknowledge that this increase in transparency will benefit the field writ large. One of the low-hanging ideas is to mirror the financial sector and create a marketplace of charitable investments. This 'marketplace' is an online place where prospective donors can peruse potential projects to support and monitor the progress of organizations in real-time.

PART 6

Ranking the Health of Nonprofits

We could start by simply modeling the inflow of philanthropic dollars to an organization. The flow of donations to a particular charity is often based on information that individuals have. With positive information about the work of a charity, more money is likely to flow in their direction. Conversely, if there is a rumor or negative information out there that individuals are privy to, it will likely negatively impact donations. We don't even need to know the granular details, it's just seeing how other donors respond and act as demonstrated by their giving.

If everyone was donating to St. Jude's Children's Research Hospital (SJCRH), we may also choose to follow this donation based on perceived increased confidence in SJCRH. Conversely, if there is an decrease in donations to another charity, say Wounded Warriors Foundation, we could view this as possible evidence of a drop in donor confidence. In both cases, this notion could be completely wrong. There could be a variety of reasons people donate more or less to a certain charity. In the aggregate though it's an intriguing concept. Imagine the ability to tap into the internet to see if your favorite charity is getting more donations or less. Download an app that shows charitable contribution flows. Again, the reason for the increase or decrease can vary and could also be due to ancillary factors completely outside of the control of your favorite charity.

Outflows or lack of funds flowing to an organization could signal they are not following through on their promise or their services or intervention just isn't working. Further it could even signal that something is amiss at the organization, for example, management upheaval or a political entanglement.

The direction donations are moving, either up or down, would be a valuable piece of information for a donor. Even if you are an existing donor and the trend is moving down, it might make you ask why gifts are decreasing. Perhaps people just aren't giving or perhaps there is another reason, say for example the CEO has told a few people that they are taking another job with another agency.

In addition, inflow and outflows could also be a forward-looking indicator of an emerging problem within a geographic area or sector. Perhaps an emerging trend in the sector is playing out to have less impact than originally estimated.

Providing such information would also help target donations and grantmaking toward a point in time when the gift is truly needed. Many nonprofits rely on end-of-year campaigns for fundraising. Traditionally tied to the end of the calendar year, these draw from the sentiment of the holidays as a time of giving and also the end of year tax deadline for making a charitable gift.

Using this data, we can show an 'order book' on in this case 'pledges' at a nonprofit. Pledges appear as line items in the financials under receivables, but again you would have to really dig deep into the data to find that information. No receivables would simply mean there were no gifts outstanding when the agency filed their audit.

The data on donations could further be bundled into a sentiment index. The index would be based on what types of organizations or what sectors are receiving the bulk of philanthropic dollars. Extrapolating further meaning from this data could tell us something about the state of the economy. What does it say if donations to the Opera are at an all-time high, for example?

Such an insight wouldn't be possible without nonprofits reporting their financial status in real-time instead of reporting it long after the fact. Sure, you can ask a nonprofit for their most recent financials or their YTD financials, but that is just a snapshot in time. It might not be a true expression of their potential or if they are on a downtrend.

This level of insight will not only benefit foundations but also nonprofits by helping to drive donations when funds are needed and immediate cash flow might be more useful.

Chapter 26

Ranking Nonprofit Performance (or lack thereof)

One of the nightmare scenarios for a donor is making a sizable donation to a charity only to find out a few weeks later, via the local news, that the CEO of said charity was embezzling funds and was dismissed by his board last week. Putting aside the criminality and the unethical nature of such a crime, it can be frustrating for a donor who just wants to have an impact to feel like they threw money out the window on a donation to their favorite cause.

Information on program success and the impact of services flows mostly to the foundation financing the activity. This information is also shared with donors and posted on the nonprofit's website. Again, this is mostly a one-way street.

Imagine if we had a dashboard with indicators updated in real-time showing the daily performance of nonprofits. We could score the performance on a nonprofit based on their program performance, operations, governance and a variety of other factors (for example: a nonprofit changes their mission from helping animals in Brazil to promoting clean water in Namibia). Green means good and red is bad. The color tells you something, but not everything. The content and ratings would be generated automatically using artificial intelligence.

Foundations can take the lead in capturing this data and develop a sector-wide movement to share performance data across the sector.

Use the new data set to show where donations and grants are going and how they are being used, *in real time*. Make the data visible to others. Allowing people to view this data and see the real time impact of an organization will help direct funds to agencies and solutions that are truly having an impact. According to a 2013 study from the Center for Effective Philanthropy, nonprofits want more insight on what foundations fund and what the results they have. Most respondents — 87 percent — indicate that foundations should be more transparent about how they assess their own performance. One nonprofit leader comments, "It is helpful to know how [foundations] measure the impact they are making so we can determine whether our work and outcomes are a match for their goals." Nonprofits also want to understand what foundations are accomplishing through their work. More than 75 percent of nonprofit leaders say they want more transparency from their foundation funders about the impact that foundations are having. One nonprofit leader comments, "It would be helpful to receive more information on the accomplishments of the foundations themselves and the initiatives they are seeing the greatest success with."

Charity Navigator will provide a rating system for nonprofit. This rating system is well known in the industry and by donors. These star ratings are used by donors and trumpeted by Executive Directors if they have a good rating. Although there has been some pushback on how useful these star ratings are and what they measure (for example, a percentage of funds going to overhead), it's a logical attempt to fill in an information void. The site provides a wealth of information designed to inform the donor.

There are some broader efforts to network and coalesce data from funders. Mainly in Candid which is formerly GuideStar. Providing data or organization information is not required nor mandated. Again, much of the data here is in aggregate or points out nonprofits working in a specific field and a specific area.

Right now this information is shared through websites, glowing social media posts and the like. The information is usually filtered through the foundation's perspective. It is rare that foundations share information on efforts that are not working.

Foundations could also use AI as a bridge between existing technology and our grants management systems (assuming we are using an electronic grants management system, some foundations still do not). Currently, grantees will upload data or a report onto a central server we can access through our grants management system.

By creating a data bridge or digital connection between the grantee and the foundation, we can then customize the data and results we obtain from the grantee. We can then further build on this connection and create system-wide connections among foundations to provide a global view of our collective impact.

Integration of AI into our existing operations, databases, software and technology tools will allow for faster and more precise due diligence, better oversight of spending, tracking and verification of outcomes, greater impact more precise donations and trackable donations. A seemingly seamless world of accountability and transparency.

CHAPTER 27

Ranking Nonprofit Program Performance

On a more individual level, we could utilize existing hardware integrated into a nonprofit program to create a new avenue of awareness on behalf of the donor. Recall the earlier example of a funder providing support for yoga classes for veterans suffering from PTSD. The veterans, most of whom probably already use a wearable, can agree to wear an fMRI sensor while practicing yoga. The data from the fMRI would be captured and interpreted for the individual participant. We can share this data, de-identified of course, with the funder, so they can see the effect on the participants in real time. In the case of the yoga class, capturing fMRI feedback provides the donor with the ability to see actual data confirming the great impact of the program.

Allowing the funder to see the actual neurological impact of class participants is a marked change in how we currently practice philanthropy which will not only validate the effectiveness of the program but lead to higher levels of engagement by the funder as they can truly see the impact of their work. In turn, this information can be shared with other funders to validate the work and encourage them to also support the program.

PART 7

Artificial Intelligence

"The question isn't will we create intelligent machines, the question is will we create beneficial intelligent machines?"

ELIEZER YUDKOWSKY

CHAPTER 28

AI for Administrative Tasks

The transformative power of AI and the fact that we are already headed down this path are reasons to harness this transformative power of AI to reimagine our work. With AI, the work of a nonprofit and foundation can be greatly enhanced. The amount of data, networks, systems and data points an AI-based system can access goes well beyond that which an individual agency, program officer at a foundation, or even a team of program officers, can realistically offer.

In its simplest uses, AI embedded into software we use daily can be used to write meeting minutes, schedule a meeting with colleagues or cull information from files. For nonprofits, we can even use AI to streamline the grant writing process. In considering the use of AI to put together a grant application, while it may appear that an applicant is side-stepping the process, consider the amount of time the nonprofit could be channeling into direct services or helping the people they are trying to help or even just focusing on other important issues. AI is a time saver and relieves the administrative and reporting burden on nonprofits. CHAT GPT can be an effective tool helping grant writer's 'fill in the blanks' on programmatic content but as we all know, the success of any program depends on other factors beyond just a well written proposal.

Nonprofits also spend a great deal of time writing grant applications to submit to funders. Those same applications, if the grant is awarded, come with written reporting requirements.

Much of the information that funders often request is duplicative. Nonprofits can save a great deal of time by formulating sections of the grant application using readily available data. ChatGPT and other systems can put together an overview of most any initiative for which a nonprofit would be seeking funding. We could embed this movement to encourage and facilitate the smart adoption of AI into the grant writing and reporting process into the Glass Pockets or similar movement.

CHAPTER 29

Case Study

In 2022, when ChatGPT first appeared, a colleague of mine and I spent the better part of Spring Break going back and forth on the ethics of using CHAT GPT for work. Me being curious, I figured I would try to use this new site to create a grant application.

The output was quite intriguing, although somewhat impersonal. Nonetheless, it showed the capability of the system and clearly demonstrated the opportunities AI can produce.

Below is the example of a grant application written by CHAT GPT. Please scroll to the bottom for my thoughts on this application.

Randy:

Prompt: Write a grant application for reducing maternal morbidity among African American women in Palm Beach County.

CHAT GPT:

I. Introduction

The high maternal morbidity rates among African American women in Palm Beach County is a significant concern that needs to be addressed urgently. According to the Centers for Disease Control and Prevention (CDC), African American women are three to four times more likely to die from pregnancy-related complications than their white counterparts. Moreover, the Palm Beach County Department of Health reports that African American women in the county have the highest rates of maternal morbidity compared to any other ethnic group. Therefore, we propose to implement an evidence-based program that focuses on reducing maternal morbidity among African American women in Palm Beach County.

II. Project Description

Our proposed program aims to reduce maternal morbidity by addressing the root causes of the disparities in maternal health outcomes for African American women in Palm Beach County. The program has three main components:

A. Community Outreach and Education

Our program will prioritize community outreach and education as a critical component of the project. The program will work with community-based organizations, healthcare providers, and other stakeholders to provide culturally responsive education, support, and resources to African American women and their families. We will focus on promoting the importance of early and regular prenatal care, identifying and addressing high-risk pregnancy factors, and educating women on warning signs and symptoms of pregnancy-related complications.

B. Doula Services

Our program will provide doula services to pregnant African American women in Palm Beach County. Doulas are trained professionals who provide physical, emotional, and informational support to women during pregnancy, childbirth, and the postpartum period. Studies have shown that doula services can significantly improve maternal and infant outcomes, particularly among African American women. Our program will work with trained and certified doulas to provide free or low-cost services to women who otherwise might not have access to this critical support.

C. Provider Training

Our program will provide training to healthcare providers on implicit bias, cultural humility, and best practices for providing care to African American women. The training will aim to help providers understand the unique challenges faced by African American women in accessing quality care and provide them with the tools they need to provide culturally responsive care.

III. Project Evaluation

We will evaluate the effectiveness of the program using a mixed-methods approach, including surveys, focus groups, and medical record reviews. Our evaluation will assess the following outcomes:

1. Increased rates of early and regular prenatal care among African American women in Palm Beach County.
2. Reduced rates of maternal morbidity among African American women in Palm Beach County.
3. Improved patient satisfaction with care received during pregnancy, childbirth, and postpartum period.

IV. Budget

Our proposed program will require a budget of $500,000. The funds will be used to cover the following costs:

A. Community Outreach and Education – $150,000

- Salaries for outreach and education staff
- Educational materials and resources
- Outreach events and community engagement activities

B. Doula Services – $200,000

- Compensation for trained and certified doulas
- Supplies and materials for doulas
- Administrative costs associated with providing doula services

C. Provider Training – $150,000

- Trainer fees and travel expenses
- Educational materials and resources
- Administrative costs associated with provider training

V. Conclusion

Our proposed program aims to address the significant disparities in maternal health outcomes for African American women in Palm Beach County. By providing education, support, and resources to women and their families, we hope to increase rates of early and regular prenatal care, reduce rates of maternal morbidity, and improve patient satisfaction with care received during pregnancy, childbirth, and postpartum period. With your support, we believe that our program will improve the health outcomes of African American women in Palm Beach County.

Back to Randy:

A couple notes on the application itself:

- The content is pretty much free of typos and reads coherently. Furthermore, it contains solid cases for support of the program it proposes and even includes the usual sections of a grant application.
- One way to make the application more relevant would be to add more localized data broken down by race and ethnicity. The Project Description section is somewhat vague and lacks overview of any partners critical to this work. Lastly it does not mention the key underlying causes; racial disparities in treatment and health care access.

AI complied this information using information already posted on the internet. Descriptions of the county, the issue (lack of access to maternal health), why it is important and what should be done, abound online. When we create a grant application, there really is no need to recreate what is already there. CHAT GPT simply searches for this information and combines it into a coherent framework, in this case a grant application.

While the process of compiling a grant application can be helpful for planning and budgeting purposes, manual, paper-based processes are outdated and inefficient. They waste paper, printer ink and other office supplies like pens and paper clips. Manual processes also cost more in human labor. Those are costs that nonprofit organizations can't afford.

Technology relies heavily on automation, which speeds things up and frees up board members' and executive directors' time for more important things like fundraising, compliance, community impact, direct service, building volunteers and other activities that increase the nonprofit's productivity.

Nonprofits generally try to reach and serve as many people as they can. Technology makes it possible to expand a nonprofit's reach in the products and services they provide, thereby increasing their capacity to serve their intended population.

The right technology helps nonprofits significantly by helping them increase their revenue. For example, <u>automation is a valuable tool for fundraising</u> and recording volunteer hours.

Many donors prefer to donate electronically, which saves leaders from having to attend meetings to formally request donations.

Even with these advancements, nonprofits, foundations and the philanthropic sector reward the same solutions and may be reticent to take risks with their funds. Funders are generally locked into a funding model guided by strict adherence to limited overhead costs. Many Americans believe that these institutions spend too much of their budgets on overhead and too little directly on programs. With this framework and with this prevailing assumption, it's hard to create the system-wide changes required to harness the power of technology, reimagine philanthropy and open the door to greater impact.

Imagine how much time we could save the nonprofit sector from researching and writing proposals. Surely, we don't think the nonprofit sector needs to research their chosen cause, do we? Doesn't it go without saying that they probably are uniquely and intimately familiar with their cause? Perhaps it would be better if we channeled this time, energy and expertise into delivering solutions.

CHAPTER 30

Case Study

Supporting the Use of Technology

About a decade ago when I was still working for a private foundation, we had an agency approach us with a grant request to purchase a piece of equipment to facilitate the delivery of Eye Movement Desensitization and Reprocessing (EMDR) therapy with their patients. EMDR therapy is an evidence-based psychotherapy designed to help individuals heal from trauma and emotional distress associated with disturbing life experiences.

EMDR therapy uses specialized magnetic resonance technology to influence the brain waves of an individual. It requires the purchase of a therapeutic device to provide the treatment. They requested $30,000 to purchase the machinery. I was quite intrigued because this was right around the time we were trying to build out our portfolio of projects that use technology to expand access to healthcare. So, I made the request to our governing committee.

They were a little hesitant to provide funding because they were not entirely familiar with this technology, nor were they familiar with the EMDR method. Furthermore, the amount of funding required for the project was not dominus. But we somehow ended up securing the funding and we made the grant to the agency to purchase the machinery.

I was quite intrigued and excited to see what the results would be for this new program. I even asked the agency if they would be willing to do a presentation to some of our other partners. I thought this project would be a great demonstration of how nonprofits can use technology to improve their outcomes and positively impact the clients they serve.

At the foundation, after a project is funding and the grantee starts with the work, we received grant reports every six months from the grantee. The reports were submitted via our online grants management system. I eagerly awaited the results of the first six months of the project, and when the report came in, I read it with excited anticipation, only to be disappointed to see that the agency had not used the piece of technology that they had purchased. They had no outcomes on the individuals to which they were providing therapy using this technology. Needless to say, I was a little disappointed, so I reached out to the agency to check in to see if there was anything we could help with to make this project work.

Once we spoke, the agency shared with me that they had staffing issues and turnover of some key staff around the time they purchased the technology. In addition, they found it hard to train their staff to use the technology, so essentially, they weren't even able to use the technology they purchased with the grant funds. On top of that, some of their clients were hesitant to use the technology.

Shame on me for waiting six months to see what the results were of this program, and shame on me for assuming the agency who requested the $30,000 for a piece of technology would know how to use it and how to train their staff to use the technology. As I learned, those assumptions proved false. Needless to say, this puts the brakes on our lofty ideals to integrate technology into all nonprofits.

My first mistake was to assume that this grant was a technical solution to a technical problem. That was incorrect. Integrating technology into the daily flow of a program is an adaptive solution to an adaptive problem. They couldn't treat their patients just by purchasing a piece of technology. They had to consider the other factors; staffing and even the willingness of their clients to try this new technology.

If our program management databases and grant management systems would have been connected and we created an expectation upfront that we would also be receiving client-level data to track the outcomes on their patients, the questions around staffing and the willingness of their clients to use the technology would have naturally emerged during the due diligence process.

Grants and programs reliant on the use of technology present a perfect opportunity to marry EMDR technology into the foundation's system of oversight and reporting. Connecting the foundation with the EMDR system and providing for the ability to see results to determine if and how patients are responding creates an immediate feedback loop between the foundation and the grantee that facilitates impact reporting and accountability. Then the foundation can also make an objective case for the program if they can see how it works and then it is changing, in this case, the brain waves of the patients that they are trying to serve.

We could solve this issue to a degree by accessing the grantee's software that they would use to capture results, but this creates another step in the due diligence process. With the capability of generative AI, we can receive this information directly in a way that is easy to understand. The nonprofit benefits as well, as they can clearly demonstrate the impact of the program. They can further benefit by focusing on the delivery of the program instead of filling out a grant report for their funder.

CHAPTER 31

AI for Philanthropic Due Diligence

Probably one of the most low-hanging fruits and a solution that could be easily developed is to use AI for due diligence. We already see AI being used effectively in other industries to conduct due diligence. The finance industry, namely, has a similar model; money comes in, people sit around a table and make decisions on where the money should go. Entire companies are already in the market providing due diligence via machine learning and by using existing AI tools. Finance has even moved farther down the continuum and replacing some of their financial advisors are being replaced by AI. Even physicians are now outmatched by the ability of AI to detect cancerous tumors. Certainly, it will only be a matter of time before Program Officers are also outmatched by the ability of AI to identify the most impactful program for a foundation to support, will they not? Program officers should strive to harness this evolving power. Systems and networks should look to catalyze this momentum to advance the field by incorporating data into the process.

Currently, Program Officers are responsible for identifying the best possible projects through a process involving fielding and analyzing requests. Inputs for this process are numerous and vary by foundation. The process is more art than science. And given the inexact science and room for subjectivity on behalf of the program officers, there is certainly room for improvement. Even disruption at scale.

One of the key roles and perhaps most critical function of the program officer is knowing a lot about an agency. Some of the information is provided in the grant application by the applicant. Other information may not be provided but can be equally critical, if not more so.

When making a decision on who to fund, foundations, usually undertake an intensive due diligence process. Currently, foundations must navigate multiple systems to conduct due diligence on grant applicants. For example, a foundation staff member must utilize:

- The IRS database, which holds the organization's 990 and the statement of nonprofit status.
- The state department which oversees charitable organizations, which verifies the agency is registered in the state.
- Another state department, which verifies the agency is registered and compliant to raise funds as a nonprofit.
- Their existing in-house database, often used for grant applications and managing grant reporting.
- The grantee's website.
- The grantee's audit report, supplied by the grantee.
- Various social media sites to monitor online presence.
- Board meeting minutes, if available.
- Visiting the site where the nonprofit provides their services in-person.

The challenges in conducting due diligence in this manner are numerous: outdated information, antiquated and clunky government databases that are difficult to access and navigate, and a lack of transparency around what is on the page versus reality. Each of these sources requires time and expertise to access, analyze, and assess whether an application should be considered for funding.

An aside: My favorite indicator is the cash reserve. If it's large, there generally should be a plan for use of the funds. If it's small, it generally may indicate a number of qualities (poor financial planning or a recent reduction in funding) that may or may not inhibit or hinder an organization's ability to carry out their mission.

Funders can utilize existing AI platforms to access information and formulate an algorithm tailored to their due diligence needs. Constructing such a system would not take much time or expense. Existing AI systems can dramatically ease the workload and the amount of time funders need to conduct due diligence.

Now go beyond this step within one foundation and imagine broader sector adaptation of generative AI used to capture the impact of foundation grantmaking. What if the individual data sets owned by foundations were connected? Grantees are already submitting reports to each of us on a regular basis. Using that data showing what works and how well it is working would allow us to build a collective intelligence greater than any one person or organization. Each foundation could query data from the sector to immediately gauge the effectiveness of a program or understand pitfalls to consider when undertaking another initiative.

Continuing with this lens of due diligence, let's also consider the information that changes hands during a board meeting. Most of the time, board meetings involve discussions around grantmaking. Who are we funding? How are they doing? Is there any trending need out there we should start to consider for our next meeting? Staff members provide insight and advice, board members listen and offer their own opinions. Minutes from these meetings are generally never shared outside the four walls of the foundation, nor is there any law that says they should be shared or made available. Conversely, publicly traded companies share meeting minutes, some even livestream their earnings calls to anyone on YouTube. Philanthropy has no such requirements.

Imagine if foundations were required to submit quarterly board minute reports (which are likely already transcribed and/or recorded) to this central server? Next imagine if we then fed this information into a large language model to create a knowledge database that foundations could access. The level of data you could cull from just a large language model. Again, one of the tenets of the foundation is privacy, but given the increased scrutiny around the corpus of philanthropic and educational organizations, maybe we need to think about opening this up and allowing other foundations or even individual donors to access the expert, highly informed insights we utilize daily.

When I see headlines about large institutions paying fines, or having sizable grants pulled, I think we need to really consider whether the level of privacy that we enjoy is a long-term benefit. Again, is privacy a feature, or a bug?

Chapter 32

Case Study Example

Sample Due Diligence Prompts

After you provide your chosen AI tool with the correct data, there are numerous prompts you can use to peruse the data:

- Who is the executive director? Are there any mentions of him/her in the media?
- What program does this agency focus on? Can you find evidence of other nonprofits implementing the same program? What makes them successful? Given the 990 and budget included with this grant application, does this agency have a reasonable likelihood of success?
- Who else is supporting this agency? Have they increased their support or decreased their support? Why?
- How quickly do the people served by this respective nonprofit benefit from any donated funds? How effective is the nonprofit?

CHAPTER 33

Case Study

One of my closest and perhaps most memorable blunders as a program officer happened when I was in the end stage of a funding recommendation to a large nonprofit agency that provided mental health services. At the time, this was an area of acute need for the community and also an area in which we at the foundation were actively pursuing new investment opportunities.

Through the course of the normal due diligence process, we examined written materials provided by the applicant. I visited their site in person and talked to the directors. I specifically remember engaging with other funders about the agency to see if they had any feedback on the agency. Nothing turned up and the program had strong potential to impact the community.

Just before the grant was to be approved by our governing board, we received a 'tip' from a colleague (one not even working in philanthropy) that we may want to tread carefully if we are thinking of supporting the agency.

The next day the newspaper reported on a real estate deal that some of the nonprofit's board members were in the process of closing. The deal would sell the assets of the nonprofit (in this case four buildings in a desirable location near downtown), dissolve operations and turn the profits over to the board members involved in the deal. Not a legal activity. We pulled the grant before any funding left our offices, but nonetheless I felt like I had egg on my face. Why didn't I look into their board members more closely, beyond a cursory examination of the board list? Surely, I would've connected the dots and decided they didn't pass the due diligence test.

Upon further research, a number of articles turned up that named a few of the board members on past real estate deals that drew legal attention.

Generally, I didn't research the background on each board member of an applicant. It would probably make sense to do this, but with multiple applicants and the likelihood of nefarious activity on behalf of a board member being somewhat low, this wasn't a priority in the due diligence process. AI offers and easy fix for this problem.

Now using freely available web tools, you can prompt for this type of information. Just by uploading the applicant's 990 and identifying the list of board members, you can then ask the AI system if there are any news articles that mention the specific members. Hopefully, the only information AI finds is related to their business and charitable work. There certainly could be some personal information that surfaces, but that may or may not be a factor in your consideration.

In this instance, asking about the board members and asking the system to search for mentions of any of them in the media would have surfaced news articles of nefarious (allegedly) real estate dealings that would give a charitable investor pause.

A huge benefit of AI is the fact that it can process a vast amount of information. You can't. As a donor or program officer, you don't have the time, resources, network, insight, or background to know everything. It doesn't matter how smart you are. It doesn't matter where you went to school. It doesn't matter how successful you were in your previous career.

If anything, you are setting yourself up to look uninformed, or worse ill-informed, in your decisions. If your job is to be informed, you better be informed.

Financial companies are already using AI for investor due diligence. By scanning board meeting minutes freely available on the web, companies can provide deep insight on company specific trends and performance. They can measure the tone and language used during the meeting to tease out clues about the board's view of the CEO performance.

Publicly traded companies are required to post board meeting minutes online along with updated financials. This same framework can be applied to the nonprofit and charitable sector. The benefits of using AI in the due diligence process are hard to discount and they include:

Enhanced Data Analysis: AI can quickly and accurately process large amounts of data, finding patterns and insights that humans might miss. This helps funders and grantmakers make better decisions.

Improved Risk Management: AI algorithms can look at past data and current market conditions to predict risks. This helps funders and grantmakers manage their portfolios and avoid potential losses.

Algorithmic Investments: Human emotion often can play an outsize role in the decision making process of funders and grantmakers. Using AI reduces the impact of human emotions and biases, leading to more consistent and impactful strategies.

Sentiment Analysis: AI can analyze news articles, social media, and other sources to understand sentiment around an agency or applicant. This is also critical even for background on the individuals managing the organization. This helps funders and grantmakers see public opinion and potential market sentiment.

Personalized Investment Strategies: AI can create investment strategies tailored to a foundation's goals, risk tolerance, and preferences. This personalized approach can improve grantmaking performance and donor satisfaction.

Cost Efficiency: By automating routine tasks and analysis, AI reduces the need for human labor, lowering operational costs for foundations and individual donors.

Nonprofits and charities do not have the same requirements as financial institutions. Requiring nonprofits to post board meeting minutes online would allow more transparency around performance and can boost investor confidence in grants and donations. Requiring foundations to post their meeting minutes online and connecting them into a publicly available generative AI system would create more transparency and increase knowledge around what works and what doesn't work.

CHAPTER 34

AI for Smarter Giving

If you want to invest your money in stocks, there is no shortage of companies, experts and advisors with opinions of how to invest your money. If you want to donate to a charity, there is a shortage of experts and advisors advising *to whom or what* you should donate. Of course, most organizations have websites with pictures and statements enticing you to donate or become involved. Some even employ fundraisers who are specifically charged with finding donors to support the organization.

But what if you can't decide where to donate your money? What if, at the end of the day, all the websites and noise about who produces the greatest social impact just serve to distract? What if all the nice pictures and smiling faces start to look the same, and eventually organizations themselves start to look the same? Where do you put that money? What if you're not Bill Gates and you can't afford 1,000+ people to assist you in identifying the most impactful use of your dollars? So many diversions and so many questions to ponder for the potential donor.

Admittedly, the application of AI to philanthropy is early yet. That said, a curious scan of existing AI capabilities in other industries can lead one to think of three ways AI can advance giving and ease the donor burden.

- AI can tell you *which agencies* support your chosen cause. It's easy to align cause with agency through generated insights from online activity.

- AI can give you an idea of *who is supporting* your chosen cause. While you may stumble across and interesting page online with what looks like a worthy cause, why not rely on the past actions of donors and perhaps even your friends for your next giving decision? With no shortage of social media posts from people showcasing their attendance at a fundraiser or posts bringing attention to recent donations, AI can easily tap into social media platforms.
- Last, you may *hone in on quality* by quickly ranking local agencies via online ratings. Guidestar and Charity Navigator actively promote their rating systems. While the basis for said ratings is continually a point of discussion, outliers on the scale will stand out in any AI generated findings.

Many of these solutions require development of new but quite simple algorithms and there are quite a few barriers, some of which are eroding as you read thanks to efforts like Guidestar's Financial SCANSM. That said, financial firms are already weaving AI into their financial advice offerings. Recognize that larger foundations employ individuals just for these very types of daily tasks. Plus, more and more, donors probably do not want to wade through multiple platforms and IRS legalese just to figure out an answer to the simple question: who is worthy of my support? More and more, new donors want faster insights and more rapid feedback. Considering the possibilities to inform donors, the application of AI to philanthropy is inevitable.

CHAPTER 35

AI Generated Large Language Models

We should push for capital investments to establish and utilize Large-Language-Models (LLMs). LLMs are currently used for data repositories and corporations, we can also develop these for nonprofits and foundations. LLMs are essentially digital databases that archive data which users can access and query for data and answers in real-time. LLMs are different from databases in that they can piece together data points in the system to generate new findings.

These are perhaps one of the most practical tools available that can help with workflow and data.

LLMs are created by plugging your data into an existing AI platform. AI systems require an API to plug into their platform. Once this process is complete, the nonprofit or charity can access the capabilities of the AI system (ChatGPT, Claude, etc.), enabling them to build AI-powered features into their work. You would simply send text prompts to the API and receive AI-generated responses, similar to how users interact with ChatGPT through OpenAI's interface. These are already widely used for customer support chatbots, content generation tools, educational assistants, creative writing aids, and many other text-based AI features. For developers, the API provides a way to leverage large language models without having to build and train their own AI systems.

The amount of information, records and data currently held within the charitable sector is immense. Actively creating an LLM, even for just your own nonprofit or foundation, would lend itself to the broader connected solution. At a minimum, it would retain the institutional knowledge now held by staff.

LLM's can be produced relatively easily. Generally, it requires the storage of existing data in an AI system. The user then teaches the system just as you would a new staff person. It is also trained by teaching the LLM the most common questions and considerations for a foundation, such as:

- Is this program a best practice?
- What program is most effective?
- What agency produced the highest impact?

Once the model is trained, it can be used by foundation staff to make decisions.

It can also be used by grant applicants. For example, potential grant applicants to the foundation could visit the foundation website and interact with the model and ask queries such as:

- What is the average size grant award?
- How likely is the foundation to fund our program?
- Has the foundation funded a program similar to mine in the past?

Notice the questions and queries are standard. Notice that many of them are currently fielded by program officers. If we connect our LLMs, we can then access the vast trove of data and knowledge held by other organizations. This data can be used to produce new insights and help plan our work.

LLMs can even be used to maintain and sustain organizational memory. Select organizations are piloting the use of AI to record phone and voice conversations of CEO's and other organizational leaders. This data can be compiled and organized into an LLM such that staff within the organization can access the records of conversations to get up to speed on projects or important matters.

CEO turnover rates in the nonprofit sector skyrocketed by 85% from 2022 to 2023, underscoring the rapidly changing landscape and the high rate of organizational memory exiting the front door. Yet, with 40% of new CEOs failing within their first 18 months, and nearly half of executive transitions faltering, proper preparation is vital. CEO transitions present numerous challenges, such as diminished staff morale and culture, loss of trust, decreased effectiveness, and negative impacts on donor/funder relationships.

According to the Florida Nonprofit Institute, 30% of CEOs in the nonprofit sector will retire in a year. They are taking decades of knowledge with them out the door. Organizations can start using AI to help with leadership transitions before they happen by creating LLMs based off data captured via phone or on zoom while the CEO is still in place. This LLM can be queried by staff and generate insights into making decisions, thus aiding in continuity during leadership changes and the eventual long-term sustainability of the nonprofit.

We can also use AI generated large language models to transform how we record, document and share community support. Mutual aid describes the efforts by community members to help one another, particularly during times of uncertainty and crisis. These acts of solidarity create a sanctuary of support that strengthens the bonds within neighborhoods, cities, and regions.

Think of the original 'Green Books' used by African Americans as a travel guide through the south. The Green Books guided the traveler to friendly establishments they could frequent along their journey. This travel guide didn't just list safe accommodations and businesses—it created a network of care, or mutual aid, that enabled survival and dignity in the face of systemic oppression.

Just yesterday while frequenting my favorite locally owned coffee shop in Minneapolis, I noticed a flyer for a hotline. Not like any normal hotline, this hotline was essentially a form of modern-day mutual aid. Established by the LGBTQIA community, the hotline would record stories of mutual aid provided by callers thereby actively 'documenting and preserving stories of LGBTQIA mutual aid' to capture the personal testimonies of both helpers and recipients.

Stories would be recorded and then collected in an online archive used to collect and categorize community support efforts. According to the flyer, the information could potentially then be presented in narrative installations in art spaces to amplify individual experiences.

Documents and historical archives are an essential and critically important way to create a legacy for minorities; legacies that could be tarnished, erased or modified to suit the voice of power. LLMs offer a modern method to sustain and share the history of mutual aid.

Using the voice recordings we are already capturing, we could create a 'mutual aid' LLM. Using this LLM, we could process and organize vast amounts of mutual aid stories, creating searchable databases of community support. We can then aggregate the data within the LLM and display real-time information about when and where help is being offered, the nature of assistance provided, and the ongoing needs within a community. You could even zoom out and see a larger map which connects data sets from other Mutual Aid Hotlines across the US to show stories at a state or national level.

This digital approach preserves the immediacy and authenticity of mutual aid stories while making them accessible to broader audiences. By creating dynamic, searchable records, we can better understand patterns of community support and demonstrate the profound impact of people helping one another.

This capability is important as it enables the network to share these stories and their impact in real-time. Network members can visit a site or download a 'mutual aid' app and see stories and events near them. You could even zoom out and see a larger map showing stories at a state or national level.

The advantage is the data on mutual aid comes into the network in real-time, with no delay in downloading the data, processing into a dashboard and uploading it back onto a website.

Lastly, since the community creates and owns the LLM, the members would truly own the archive. It would not be hosted on a separate website or sold by a corporate book seller who could label the contents a threat and cease distribution. While a minor concern, with efforts to normalize increasingly hostile political retribution of late, it is something to consider.

CHAPTER 36

Case Study

Amazon's LLM

Let's look at Amazon as a possible example of an entity with data and skill to create this mechanism. Donations through Amazon Smile surpassed $215 million since the inception of the concept in 2013. These donations through Amazon Smile take place in real-time. According to their Super Bowl commercial(s), Amazon can tell us the likelihood of a receiver catching a touchdown pass. Why can't they tell us the amount or flow of charitable investment toward a given organization?

Why can't we hop onto Amazon's site and see the flow of donations and where they are headed? A substantial or marked increase or decrease to an organization could provide a great deal of data for the charitable donor.

Publicly funded hospitals receive favored tax status for the community benefit they provide. These same hospitals are required to disclose a 'public benefit report' to show how they are disbursing their community benefit. Why don't we have the same transparency standards for large donations? This could help others make decisions on where to channel their money. By aggregating the amount of donations by agency or by sector and allowing us to see where the donations are going and when they are moving, we could go a long way to creating more transparency and knowledge for the donor. Such data could be displayed and communicated using very high-level numbers and a format that many are already familiar with. Instead, we now only know of donations after they are announced. This delay in information can be a disservice to a cause. If donors were aware of where money was flowing and to whom, they may be more likely to 'follow' the money and donate as well.

PART 8

Sensors

This chapter will explore emerging use cases for existing hardware created by sensors. Generally, we think of these as technology used at the individual level to track data. But when we combine the individual data across communities and sectors, new insights emerge that lead to more targeted interventions. Using Generative AI, we can weave this information into existing data systems to create new systems of accountability that benefit the people most impacted directly affected and create a complete picture that offers new insight and creates new opportunities to reimagine philanthropy.

Almost all of us currently use or have some type of sensor attached to our body or in our pockets. These sensors can provide a myriad of data points, where you are, where you spend your money and even your current health status.

We can take the data from sensors to the next logical step: using sensors data from human beings and transmitting the data to others to facilitate decision making. It sounds Orwellian, doesn't it? Well, look at your phone and the watch on your wrist. Maybe even the ring you are wearing or the glasses you have on your face. In some if not most cases, this information is already being continuously monitored and shared into the database of a second party corporation such as Garmin, Ray Ban, Facebook, etc. What they do with the data is not a question that is asked very frequently nor is it one which elicits much of a response other than the generic 'we utilize user information to build more engaging user systems,' whatever that means.

Walk into a stadium and your face is scanned. Going through security at an airport and your gait is analyzed by sensors in the ceiling. The end game here is security, which we have a very high acceptance rate for in our society. Sensors can obtain real-time information about an individual's physical and mental state. Apps and data sensors in mobile phones can be used to detect Parkinson's Disease by detecting tremors and measuring the gait of the wearer. We can track the rate of disease progression by monitoring whether the gait changes.

Again, corporations already access this data and use it for their own purposes (i.e. make more money for shareholders). The health care system accesses this data but not always in real-time. The data is available, it is not being shared or packaged in a meaningful way to further the work of nonprofits and foundations.

Imagine if charities captured data driven by technology and AI on their clients. Put aside the privacy argument for a moment. What if that information was then shared with the donor to provide real-time updates on how their money is being used, who is benefiting and the degree to which they are benefiting. We could aggregate this information onto a dashboard of all the clients served by the nonprofit. The dashboard would show where the person is on the continuum of care and how much farther they need to go in order to be healthy. We could see the impact of the organization's work in real-time.

This data could also be used to judge the performance of the nonprofit. If you are a person who receives therapy from a nonprofit, we could use these sensors to monitor and construct an image of how the service is helping you and whether or not their interventions are working. Say for example, if 80% of a mental health organization's clients are in bed 90% of the time, you could probably get a sense that their therapeutic intervention may not be working. Maybe the watch on their sensor is showing an acutely high rate of stress.

Conversely, what if another nonprofit can show more regular stress levels among their clients while they are attending an in-person meditation session at their facility? What if we can see a real-time readout showing us that an intervention is working on a human being? Wouldn't this be worth it to the system at large that is trying to alleviate pain and suffering in some form or shape among individuals and society at large?

The agency that provides yoga classes for veterans suffering from PTSD could plug the body and neurological sensors worn by the vets into a central database that creates a physiological and mental image of each participant. People who support or want to support the agency could monitor the changes of the program participants in real-time. Organizations producing a positive impact on their program participants could verify their program works and eventually obtain more funding.

Again, corporations already access this data and use if for their own purposes (i.e. make more money for shareholders). The health care system accesses this data but not always in real-time. The data is available it is not being shared or packaged in a meaningful way.

Chapter 37

Sensors

Three Examples

Using three types of sensors will enhance our ability to track programs and determine the impact of our grantmaking. These sensors include retinal sensors, neuro sensors and facial recognition sensors.

In addition to verifying identity, retinal sensors can be used as an additional layer of security to protect the donor and the nonprofit. Each party can use a retinal sensor to verify that the person on the other end of the phone or sending you a donation is an actual person. For the donor, this can help when we are considering a donation and are not entirely familiar with the nonprofit. The nonprofit can use retinal sensors to verify their activities and provide these to the donor to demonstrate their activities.

Retinal sensors can also be used to ascertain the health of an individual and measure their response to stimuli. In the earlier example of the individual receiving therapy, we can capture this information and then make it visible to the donor as verification that their funds are having an impact. If we are worried about privacy, we can de-identify the data but maintain an aggregate count of how many people are benefiting. Our phones already have sensors that can be upgraded to read our retina. We can also attach a sensor to a mirror in a patient's home to obtain data on their current health state.

Facial recognition technology offers numerous opportunities in therapy. It can gauge how someone is feeling during sessions, providing immediate feedback on the effectiveness of the therapy. This technology can also measure the success of specific programs or interventions. By using artificial intelligence, facial recognition sensors can mine data to develop treatment protocols based on likely diagnoses. Google has partnered with a network of Catholic organization clinics to capture data on Google Cloud. This data is then mined using AI to create treatment protocols. The key to this process is data capture; without data, there is nothing to mine. A telemedicine platform for therapy can incorporate facial recognition software to discern the thoughts and feelings of individuals. This can happen based on various facial reading sensors. Based on the subtle movements of the facial muscles, this technology can diagnose conditions, track the impact of treatment and therapy, and maintain patient records. Information on the success of treatment can then be provided back to the foundation or donor in real time, so they can monitor the impact of their donation instantly.

Neuro sensors, or wireless brain sensors, are devices designed to detect, measure, and interpret neural activities within the brain and nervous system. Using implanted sensors, they capture brain activity from neurons while the person goes about their day. These signals can connect with an external interface that we can monitor, allowing for real-time feedback on how someone responds to a stimulus, therapy for example. Neuro sensors are still an emerging technology and one that we are not quite there yet in terms of utilization. Nevertheless, their adoption will move the needle on program monitoring and our ability to measure the true impact of a program on an individual.

Neuro sensors allow us to use neurological responses, perceptions and feedback obtained in real-time, to formulate new insights to drive impact. We can match our cellular and neurological response in any given moment with our chosen cause. For humans, viewing a cute puppy, seeing or hearing someone we love or our looking at our favorite painting elicits a neurological response. We can't sense this response, nor do we know it is happening, but if you are alive, it's there. Advancements in technology now allow us to view the chemical and biological responses and activities that take place in our brain. Neuro sensors would elevate our ability to target our programmatic interventions and understand what is working and what is not. One area where these may prove useful is the field of equine therapy. If you are a pet lover like me, you know the feeling you get when you see an animal. It's that feeling that draws us closer for a pet, or a tussle of the ears. That feeling is your brain's response to the stimulus (in this case the pet), or in scientific terms, the stimulus response. Imagine if we could use neuro sensors to gauge the impact of equine therapy on an individual. Equine therapy has shown itself to be especially helpful to calm anxiety and to have a positive impact on mental health outcomes. The impact of the therapy is usually measured by changes the program participant reports, such as increased confidence and self-esteem.

Utilizing neuro sensors for the same program would allow us to measure brain function in real-time. Using an interface connected to the neuro sensor output, we could then transmit that data onto a dashboard that the program manager can read to instantly gauge the impact of the therapy. To maximize utility, take this one step further and make the information available to a donor and you can literally see the impact of your money in real time.

CHAPTER 38

Case Study

With a broad adoption and integration of technology into our work, we can improve medical treatment for individuals and increase compliance and accountability of industry. This is especially true in the case of polluters. According to the World Health Organization, air pollution is the single biggest environmental health risk worldwide, killing 7 million people prematurely each year. This is an issue affecting community here in the US as well. Neighborhoods built next to industrial areas or factories that pollute the air and are disproportionately negatively impacted by pollution are often referred to as **fence-line communities**. These communities are located near industrial plants and are exposed to higher levels of pollution, which can have significant impacts on the health and living conditions of residents.

Additionally, such areas are sometimes called **sacrifice zones**, as they bear the brunt of industrial pollution and environmental degradation. Residents in these zones often face disproportionate health risks, including respiratory diseases, cardiovascular diseases, and cancers.

Sensors are already used to monitor pollutants in the air. The government installs them and tracks them. County Health Departments and Industry often install air sensors around fence-line communities to monitor pollution. You can visit a website and view the readout from the sensor. However, the sensors are often not working, or the data is presented in a format that is only clear to scientists.

It would be very useful to use sensors in areas with known cases of asthma and present the data in a clear format which people can view in real-time. This data would be especially useful for the people living close to these areas.

In Western Palm Beach County, residents live under the cloud of pollutants generated by the sugar industry. Depending on who you ask, there are varying answers to the question of why there is such a higher incidence of asthma among the people of this community. It's not known or tracked how many pollutants come into the air in a way that is accessible to the residents of this community. Poor health conditions, especially asthmas, are particularly high. Most signs would point to the activity by the sugar industry, although the corporate interests at play here will tell you otherwise.

In the '20's, the sugar and agricultural industries thrived on forced servitude and slave labor imported from the Caribbean, even after the abolition of slavery. Alec Wilkinson detailed the experience in his book ' .' In some ways the third world conditions in the western area of Palm Beach County persist to this day, with farm worker immigrants living in substandard housing while helplessly exposed to toxic air and working conditions. But in the '20's the conditions were especially harsh.

According to the Sierra Club, 'Pre-harvest sugar field burning is a toxic and outdated harvesting practice that takes place every year from October through May over the approximately 400,000 acres of sugarcane fields in and around the Everglades Agricultural Area (EAA). Farmers burn sugarcane crops before harvest to remove the leaves and tops of the sugarcane plant leaving only the sugar-bearing stalk to be harvested. This unnecessary harvesting practice negatively impacts the health, quality of life, and economic opportunity of residents living in and around the EAA.

Discriminatory burn regulations based on wind direction ensure more affluent communities to the east are spared when the wind blows their way, while residents in and around the Glades (predominantly lower-income communities of color) remain unprotected from the smoke and ash — when the wind blows toward them, burning permits are granted.

One way this pollution manifests itself is in vastly different numbers on a page. For example, hospitalizations for Chronic Lower Respiratory Disease (CLRD) among African Americans in Palm Beach County are nearly double the rate among whites. CLRD stems from negative environmental causes, substandard housing conditions and poor air quality.'

Much of the western region of Palm Beach County where Belle Glade lies is oftentimes blanketed in soot and smoke from the burning of sugar cane. This is what is called a 'health inequity.' Health inequity is described as a difference or disparity in health that is systematically associated with social advantages or disadvantages. Even though there is an obvious factor in the alarming rates of CLRD among the residents of this area, the powers that be do not acknowledge or seek to change the practice.

It can be difficult to prove causation when it comes to health. In addition, industries polluting the environment are not readily providing data on pollutants emitted from their plants. Although if you drive west in Palm Beach County out toward the lake on a winter day when the cane is burning, it doesn't take a genius to realize that the black smoke emanating from the fields and lingering in the air is probably not good for your respiratory health.

If we can mount sensors on streetlights and allow the residents of the communities a way to monitor and flag when pollution is above dangerous levels, we shift the power to the community and away from the convoluted (well-meaning nonetheless) local bureaucracy. If the sensors are installed and monitored by people who live in the communities, we can bring more accountability and visibility to this problem.

The government may monitor the pollutants, which they do currently, but holding the perpetrator accountable is rare. Too many large corporate interests are at play here to be held accountable. By seamlessly providing information in real-time directly to the residents using a sensor, we put the power back into the people's hands. They do not have to rely on large well-meaning government institutions to collect data and air quality samples for them. Even if they do collect samples the data must be obtained from the government. Sensors which can be used by the residents would be more impactful in creating a system that prioritizes the health of the people and puts the power in their hands to bring meaningful change. Sensors can help protect residents from pollution. They can be used in real time to monitor the amount of pollution generated, feed that data into a financial system that creates an immediate accountability mechanism in terms of a fine (not delayed or adjudicated by a court or determined by a panel) the industry has to pay immediately.

This level of insight and transparency empowers the people and brings about more equitable environmental protection. This happens internationally, in Sweden for example, researchers at Chalmers University of Technology in Sweden are mounting small sensors on streetlights to monitor air pollution.

Furthermore, if we are channeling taxpayer dollars and/or charitable donations to this area, we can also see whether or not our efforts are paying off. If they are not having an impact, the polluter could be held accountable in real-time by assigning a financial penalty that is accessed immediately on the industry.

Monetary penalties to polluters are issued, if at all, years after the fact. The money usually goes first to a government agency, who may eventually distribute the money in a way they see fit, usually in the form of mitigation projects back to an organization or a developer. Again, community members are bypassed and are allowed minimal impact in terms of a meaningful use for the monetary penalty.

Using seamless data transfer capability of AI, the pollution data could be monitored by regulatory agencies which could in turn instantly issue penalties against the polluter.

Instead of requiring community members directly affected by pollution to log onto a website to gather data on whether or not their air is polluted, they can receive updated data in real-time, such as notifications via a phone app. We can connect pollution data with the data systems and electronic medical record systems at the local health care providers and physicians so when a patient presents with acute asthma, the physician could simultaneously receive the data to make a treatment decision. We could charge the polluter for the physician's visit, instead of sending the bill to an individual (who isn't a fault) or an insurer (who probably is based in another state).

CHAPTER 39

Sensors to Monitor Program Outputs

The use of sensors way be especially useful as a way to monitor how well a nonprofit is performing. Sensors can also be used to track outputs. These outputs can be aggregated and reported in a way that benefits the donor. We can use sensors to obtain data that is then transmitted back in real-time back to a centrally accessible database. A great example of a group using sensors to monitor their charitable output and activity is the international organization 'Charity Water.'

In the case of Charity Water, donors can literally monitor the output of a well on the Charity Water website. Charity Water uses sensors to monitor wells. Connected to the other sensors via Amazon web services, the sensors aggregate numbers of amount of water, flow, etc. This serves to precisely measure the impact of their program, brings direct transparency to the donor on the impact of their donation. If you are a donor and want to support their work, you can even identify which well you would like to support.

According to their website, because of their proximity 'the only way for governments and NGOs to monitor rural water systems was to visit them. We realized we needed to monitor our wells from afar. But no tool existed. So we made one.' The sensors not only save time and money but allow Charity Water to compile an online performance dashboard for each well. The dashboard shows key indicators such as Liters per hour, GPS, Temperature, Device performance and Humidity.

The Charity Water donor can also designate their 'donation' to a specific well in a specific location. Because of the way they have integrated sensors into their hardware, information flows quickly to the donor. Dollars flow quickly to the project (in this case the well). Impact is aggregated using technology and displayed in a way that is meaningful. Donors can literally see the impact of their gift in a completely transparent manner. It's a very innovative yet simple way to use sensors to show the impact of our work in real-time as opposed to posting updates on a website of sending a report to a donor or grantee.

CHAPTER 40

Sensors

Program Oversight and Monitoring

Medicaid fraud is a big business. Most anyone probably agrees that we need to reduce Medicaid fraud. Most of the attention in preventing fraud focuses on the service providers. Within the home health industry, mobile sensors are used to track employees and reduce Medicaid fraud. By tapping into the sensor built into their cell phone, supervisors can determine exactly how much time a home health aide worker spends in a patient's home providing care. They can track how much time they spend traveling from one client to the next.

Even though this granular information is available, there remains the false narrative among many politicians that people are lazy and sitting at home getting Medicaid benefits. Politicians cite a survey that says people would rather do nothing. Even though state efforts to require a work requirement have not increased employment. We recently saw these efforts play out successfully when Medicaid was cut in the budget bill of 2025. Efforts to push this false narrative have paid off and will likely continue to do so into the future, if we find it difficult to counter the narrative.

The ability to track and read information provided by cell phones is already in use. We are using it on a very limited basis within the nonprofit sector, if at all. For one, the sector relies a great deal on trust. The other reason is that funders generally don't require this level of regularly updated, real-time assurance that their funds are being used for the intended purpose.

We may need to consider including data sensors embedded in buildings or rooms to verify human activity. We could also start aggregating geocoded data from cell phones to verify that people were at a site instead of supporting an empty shell of a building.

This is an invasion of privacy, and we shouldn't burden people in need any further. I agree that it is Orwellian. However, I think the tradeoffs are worth it. If people continue to push a narrative the nonprofits conduct wide-scale fraud and they can't be trusted, where is this taking the sector?

PART 9

Blockchain

We could write an entire book about AI exploring how we can use it. But the true transformation in the way we approach our work includes other technology. These systems are only now seeping into public consciousness and do bear exploring to truly reimagine the nonprofit and philanthropic sectors.

The next mechanism we will explore and another essential tool in our journey toward reimagining philanthropy is the blockchain. The blockchain is the underlying technology upon which Bitcoin is built. The blockchain is essentially a public ledger that records all transactions and can be viewed by the public to verify them. The blockchain was formed on the concept of decentralized ownership and decentralized control to maintain a decentralized record of transactions. It made it possible to move units of value between computers. Anyone with access to the internet can view the ledger and can make decisions based on what the ledger tells us.

Within the nonprofit sector, migrating financial transactions onto the blockchain would save time and money. Currently, traditional financial intermediaries take a percentage cut of each financial transaction. They also hold the money for 3 or 4 days while they wait for it to clear. Banks will tell you this is a critical function and one that must be preserved. I'm not sure about you but the thought of money that I earned and deposited into my account, only to have to wait 3 or 4 days for them to 'clear' the funds is a bit frustrating.

The blockchain allows us to break free from the current hold of digital companies. Most of the companies we turn to for data management, donor management and financial services are owned by larger corporate entities. While they provide a valuable service, they also take our data and turn around and sell it to advertisers and marketers. Once we upload our data into these systems, it really isn't ours any longer. We relinquish control of our information.

We can use the blockchain to confirm transactions between donors, nonprofits, foundations and clients. By creating a record of these transactions on the blockchain, we can determine the timing of a donation or grant, when the funds were accessed by the nonprofit and whether or not the funds were utilized. Without a delay in receipt of funds, as is currently the case when sending money through the traditional banking network, nonprofits could immediately put grant funds to use. In addition, once the funds are transmitted beyond the nonprofit, we can further track their use and identify when they are put to good use.

Unfortunately, grant funds are not always put to immediate use. Anecdotally, I have heard of foundations making gifts to a nonprofit, only for the nonprofit to sit on the funds for months while they try to determine what to do with the money. While well meaning, sitting on the funds and delaying their transmission can, in some instances, delay the impact which the donor originally intended. Certainly, if someone is donating to a capital campaign this delay in using the funds wouldn't be unusual, but in other instances, it would be helpful to know the funds were put to use immediately.

The blockchain is already being used by the government. In limited cases, before their funding was reduced to nothing, USAID was using blockchain technology to improve the tracking and reliability of foreign aid distribution. Even outside the purview and largess of USAID, organizations of any size can still use the blockchain to:

- track aid distribution, ensuring security, transparency, and traceability.

- create an immutable digital ledger to track dollar spending and charity shipments, facilitating easier identification of funding misuse and fraud.
- decrease financial administration costs while enhancing rapid partner payments in the field.

Since the advent of the blockchain, there are now additional 'layers' that can be built on top of the existing chain. These are called 'layer 2' solutions. Hundreds of these now exist, and many of them allow for transactions on top of the blockchain. They maintain the decentralization principles of blockchain but enable developers to develop applications useful in finance and healthcare, to name a few. Think of them as an app on your phone, but these are 'apps' built on the blockchain. International aid organizations are already using these solutions to move money across the world. They offer privacy and security much greater than the traditional banking system.

CHAPTER 41

Disaggregating Power

The financial system in the United States has a history of discrimination and predatory practices targeting minorities. See the earlier discussion recounting the fraudulent mortgage loan debacle used by the financial sector to target low-income homeowners.

For these reasons and more, many are coming to see and understand the value of Bitcoin as extending beyond just a medium of value exchange. Increasingly, Bitcoin, and more specifically the blockchain upon which it is derived, is emerging as source of power disaggregation. Many users and early adopters were intrigued by the notion of a decentralized, equitable power structure that moves finances and records transactions.

The availability and use of Bitcoin is a means to create an equitable and non-discriminatory financial system. With no intermediary or central decision-making authority, people can obtain Bitcoin for their own personal use. This leads to peer-to-peer transactions that become ubiquitous.

Bitcoin makes sense for many people in the world. Examples of governments that debase their currency or just outright move money out of a bank into government coffers abound. Citizens of these countries, understandably, flock to Bitcoin as a preferred medium of value exchange. Financial power moves toward the people and away from the central authority.

Therefore, Bitcoin is an equitable means to transmit power via a medium of exchange. Blockchain technology via access to the public ledger upon which it is built can show you exactly where your dollars went and who they helped. We can use this additional transparency to point to the impact of charitable donations.

Currently, if we want to send a dollar to a friends or make a donation to an organization, we have to send it through a third-party (bank or Venmo) that will take 1-2% of my money to enable this transaction. We can't see how others are channeling their own donation because the banks are not transparent with their transactions.

Bitcoin, the first currency on the blockchain, can be sent directly to another person without needing a third party to verify the transaction. This is a key difference and one of the main advantages (and in cases of fraud, hacking and cybercrime one of the key disadvantages) of Bitcoin and digital assets.

Using a similar public ledger to direct funds to nonprofits instantly provides a transparent mechanism that anyone can access to see where the funds are going.

Why would you want to transmit money to someone secretly? Transmit donations directly to an individual via the blockchain becomes useful when we start channeling donations to difficult-to-reach populations or those who survive in secrecy, such as refugees or victims of abuse or human trafficking.

Usually, and not without just cause, people equate bitcoin and the blockchain with activities like this with illicit activity. There have been cases where this is true, and security continues to be an ongoing challenge in Web 3.0. However, there are other cases where secrecy can mean freedom.

CHAPTER 42

Case Study

Take the case of a group in Florida which has enabled crypto donations to their site and facilitates transfers to individuals they support. Why would they want to allow donations using cryptocurrency? Because this organization helps victims of human trafficking break out of their abusive system and away from their captors. Individuals being trafficked for illicit activities are not allowed to keep money for themselves. For them, money can be used as an escape route and a means to break free. By giving them the ability to receive currency on their phone, they can then use it in exchange for rent, etc., as long as they find someone willing to accept digital currency. This is one example of the 'freeing' mechanism of digital currency.

Why can't they use Venmo? Venmo and other payment services require a third-party intermediary, such as a credit card or a bank. They usually require you to prove your identity via a state-issued identification. Victims of human trafficking generally do not have bank accounts or credit cards or identification because their captors will not allow them this freedom.

In addition, by receiving cryptocurrency, the donor can maintain anonymity, if they choose. Until 2023, no home address was required to open an account on Coinbase or other digital wallets. Know Your Customer (KYC) laws were adopted by the cryptocurrency industry only recently.

CHAPTER 43

Blockchain Based Foundations

Once way to increase transparency, completely shift the power dynamic and disaggregate the centralized nature of giving while still maintaining a high-level of oversight is to construct a Decentralized Autonomous Organization (DAO).

A DAO is a blockchain-based organizational structure in which there is no central decision-making authority. Instead, governance decisions including the use and allocation of funds are made through distributed and autonomous mechanisms.

These are wholly internet-based businesses collectively owned and managed by its members. There is no incorporation with the state or federal government. Anyone can join. Anyone can vote. Anyone can table a recommendation for what the DAO should do.

DAOs utilize the blockchain to create collective decision-making ability on behalf of nonprofits, leaders and community partners. This distributes giving power to individuals on the forefront of whatever cause they are supporting.

Depending on the DAO, blockchain technology can also be used to create governance 'tokens' that participants in a network can use to exercise input and control over the funds of an organization. These are 'layer-two solutions' based on Ethereum or other chains such as Solana or Stellar Lumens.

The DAO increases transparency and eliminates the administrative body in the middle of the transaction. Eliminating the administrative body results in clear and transparent work with real-time information flows. There is no reporting to the IRS on a 990 that a donor can view 18 months later.

This folds the donor directly into decision making process that is entirely democratic. It is even faceless in that no one knows who owns which tokens, therefore you won't know who is voting. It is a system wholly controlled by its grantees and donors.

The Big Green DAO has been one of the most notable DAOs to emerge. The Big Green DAO labels their effort as 'an experiment in democratizing and decentralizing our grantmaking.' DAOs made headlines at the start of the Ukraine war when individuals wanted to send donations to the war in Ukraine. The donations were for various causes, but the funds went directly to an individual.

As usual regulations and laws have not caught up with this new development. Nonetheless, there are legal considerations around the use of this structure. According to Jeremy Coffey, the Founder of Mission Driven Law, 'laws may vary by state and in some instances may be non-existent.'

While ownership is decentralized, responsibility is also decentralized. There are group mechanisms to guide implementation of the project. Currently most projects use Discord to establish a communication channel for the project. As opposed to regular communication processes, sending out emails, requesting meetings, prioritizing discussions, that emanate from a central point or a central person or group, using a common open communication platform results in the inclusion of everyone in the work of the project. While this does result in more transparency and joint ownership, this can obviously lead to other problems during implementation.

Groups can use this structure to create their own tokens. They own the token. Artists are using this method to make a profit. For artists, part of it is removing the middleman and truly owning their work.

If you buy the currency, you know exactly where the money is going. It is not transmitted through the traditional financial system. Our President even has his own token. Owners of the tokens receive clear benefits, namely an audience with one of the most powerful people in the world. A benefit which people who do not own the token do not enjoy.

CHAPTER 44

Stablecoins

Stablecoins offer an easy entry onto the blockchain. They are connected to traditional currency, usually in a 1:1 ratio. Establishing a stablecoin allows an entity to create their own investment assets which can be bought and sold. Stablecoins can be used as investment vehicles to build monetary capital, while bypassing traditional sources of capital. The general advantage is one of freedom and control of fees, which are usually set by a corporate banking entity.

Many banks and financial institutions now allow their clients to purchase stablecoins. The most common is USDC, a dollar-backed stablecoin. USDC can be purchased using dollars, then converted into any number of other digital currencies or tokens.

The field of finance is moving in this direction. According to the Wall Street Journal, stablecoins function as digital dollars in crypto markets, and are currently used to store cash or purchase other tokens. Banks are in discussions to create a stablecoin customers can use in lieu of dollars or hard currency. Nonprofits and foundations should begin to migrate a portion of their finances onto the blockchain via stablecoins. With the advent of publicly traded companies who manage these transactions, this is an easy transaction.

Now with recent legislation and a congressional body intent on leading the crypto revolution, more opportunities will continue to emerge to leverage blockchain technology in ways we didn't think of previously.

The GENIUS Act (Guiding and Establishing National Innovation for U.S. Stablecoins Act) is landmark cryptocurrency legislation passed in the summer of 2025 that establishes the first comprehensive federal regulatory framework for stablecoins in the United States. As we will see later, stablecoins are already in use within global philanthropy.

Stablecoins provide an ease of transferring money to groups and individuals across national boundaries which make their use essential for our evolution. By re-orienting the way we think and manage our finances, we can effectively re-position the sector to take advantage of these types of legislative movements.

Within the nonprofit and philanthropic world, social impact tokens describe a group of digital tokens used on a blockchain with the specific goal of unlocking investments for projects with positive social and environmental impacts in support of specific pre-defined goals. Once created using the Ethereum blockchain, these tokens can be used by investors to support or grant dollars to a charity. The tokens allow for additional confirmation and security on behalf of the donor to ensure their funds are going directly to the charity of their choice.

CHAPTER 45

Ethereum

Ethereum is the second largest cryptocurrency behind Bitcoin. It has a blockchain or chain separate from the bitcoin network. Think of the Ethereum blockchain as the internet, but the Ethereum 'chain' is separate and independent of the existing internet.

Created by Vitalik Buterin in 2013, this separate internet is a network that allows for the creation and management of decentralized applications. But first, what is a centralized network? Think of Dropbox, Docu-sign or your preferred grants and/or donor management system as a centralized network. They are all useful and essential software systems created and owned by a third party. While you pay to use the system, you don't own any of the data that you provide the system. You can't do anything with the system other than utilize the capabilities established and provided by the company that owns the systems. Furthermore, and perhaps most importantly for our purposes, each system is isolated from one another. They rarely if ever communicate and, unless you are a fluent coder, it is difficult if not impossible to migrate and exchange data across these systems let alone create new findings or analysis for the data contained in each respective system.

Like the blockchain protocol underlying the Bitcoin network, the Ethereum software system is de-centralized in that no one entity owns or governs the system. Currently, the banking system is owned and managed by the banking industry. Rules and regulations govern the use of the system. Importantly, the rules and regulations within the Ethereum system are governed by the users of the system, a concept known as 'proof of work.'

In my opinion, the more important point of Ethereum is not its ability to serve as a transfer of value, like Bitcoin, but the fact that Ethereum is essentially a software system or network upon which we can establish and create new applications leveraging the 'proof of work' concept inherent in the network.

For example, we can create a contract for grantmaking on the Ethereum network. Currently, we write a contract, and our grantee signs the contract. The arbiter or ultimate oversight for the contract is the court of law. The court of law is disconnected from the grantee's ability to meet their contractual requirements. The courts do not ensure that the grantee is compliant or in fact meets their outcomes. Ultimately, this is incumbent upon the foundation. We are the judge of contract performance. Foundation staff can attest how much time they spend reading reports from grantees.

The Ethereum network allows for numerous advantages over our current disjointed system of oversight. For example, we can create 'assurance contracts' on the network to ascertain contractual performance. These contracts rely on the collective agreement of multiple parties to participate in a joint activity, such as a collective impact project. Collective impact projects require multiple organizations, each with their own unique input, knowledge, skill or activity they provide for the project. Assurance contracts make sure that a project only happens if there are enough participants. Binding collective impact projects into an assurance contract created on the Ethereum network would protect the funder by ensuring that all the organizations participating effectively coordinate their activities.

Perhaps a more concrete example is use of the same mechanism to provide assurance for a capital campaign. Capital campaigns require raising sizable amounts of money. Multiple parties are usually involved on the building side. A multitude of activities take place to ensure the building is completed: materials are purchased, plans are approved, donations from other foundations are solicited and received.

Currently, we contract with a grantee to provide a service or in this case build a building. Our verification method involves calling, emailing or even visiting the grantee to verify they are in fact meeting their contractual requirements. Most do and there is no urgency around ensuring that grantees will meet their requirements, but the system of verification that can occur on the Ethereum network greatly reduces the time we spend on contractual oversight. If the contract is on the network, we can verify that the next round of funding was raised, or that the Certificate of Occupancy was issued by the city. We can even tell when it happened.

In this case we can validate activities using existing technology (scanner and bar codes on palettes of food, for example) which are electronically connected to the network. We essentially bypass the human and administrative elements of validation and create a more seamless and transparent validation system that occurs in real-time and allows us to see activities in real-time.

We can also use the network to reward contributors to programs or services. Using techniques embedded into the chain, and separately verified by other participants in the network, we can distribute funds to organizations outside of the normal funding stream defined by the funder. This concept, known as 'Deep Funding', further encourages the development of solutions which can be applied to enhance the impact of a given program or service.

For example, we all know the value and impact of volunteer human capital. Using this same 'assurance' system and connecting a token which provides additional funding to other elements of the project that positively impact and contribute to the performance of the program is a way to incentive and reward volunteers.

Do we need this? Not necessarily, as all humans are altruistic and want to do the right thing. But with new chains it's worth exploring ways to recognize and contribute to the volunteers and people who provide care to a neighbor, picking up trash along the road just because they don't like litter or paying off the repair bill for a homeless person at the bike shop.

CHAPTER 46

New Frameworks

Migrating compliance and financial operations onto a central ledger requires relinquishing control. Sending money directly to an individual without an intermediary is a loss of control. Boards and trustees of nonprofits and foundations are generally not looking to give up control and would counter this argument by citing their role as fiduciary. Furthermore, this degree of financial transparency may be uncomfortable. Granted, that is an important role, but this new type of framework which supports a digital transformation of the sector requires some people to relinquish control.

Understandably so, trustees can be a little uneasy and when you say you are going to let the people decide what they want to fund and how they want to fund it. Oftentimes it truly is their money, but that's not always the case.

That individuals and organizations will use a bank and a central, trusted financial institution are just an assumption for everyone. Transparency around where money went, when it was used and how it was used would enlighten the field and provide critical information. Many donors and individuals spend a lot of time trying to figure out, examining and auditing organizations to see how money was spent. Entire industries and consulting companies have been created to monitor and provide feedback on the operations of a nonprofit organization and the impact of a foundation's grantmaking.

The 2020 article by Donna Daniels and Kelley Buhles of RSA Social Finance expanding on the number of ways to shift power away from the central authority of the foundation into the hands of those most impacted by issues or challenges and 'trusting them to make decisions based on their expertise.' This requires moving past decades of thinking and the administrative frameworks we have taken for granted.

This transformation will take time, but the benefits outweigh the risks. The creation of a currency based on a blockchain (bitcoin) created a new world. Foundations and donors can consider constructing and utilizing an open decentralized universe online that allows transparency and visibility around where funds are going and whether or not they created positive social impact. As stated earlier, if everyone knows where the money is going, how and whether it is being used and what the impact is of the funds we provide, we can foster a new realm of broad and highly scalable giving solutions, to create more impact.

Broad adoption of this technology and the thinking behind it will aid the nonprofit and charitable sectors writ large. The blockchain will contribute to a higher level of transparency but also greater knowledge that donors can use to make charitable decisions.

Reduced overhead spending due to improved operational efficiencies and 'disintermediation' through blockchain technologies (e.g., direct donor to beneficiary giving) could help restore faith in charitable giving among skeptical givers. This in turn could lead to increased philanthropic engagement and a rise in overall giving.

Admittedly, many of these concepts are not ready for primetime. Fraud and 'rug pulls' in the cryptocurrency world are a constant threat. More recent challenges around the use of 'meme coins' by elected officials and their family members only add another layer of hypocrisy to this subject which leads to further questions around legitimacy. Nonetheless, the concept of central ownership and a central verification mechanism that the blockchain offers could go a long way toward a more transparent donation and a more accountable donor.

So while the technology may not exactly be ready for mass adoption and our sector may not be ready to embrace these movements, it is safe to say that since these use existing technology and are continuing to develop new use cases daily, reimagining our work takes a willingness to at least be aware of them, watch them closely and consider how to incrementally adopt and integrate them into our work.

PART 10

Essential Elements Required to Reimagine Philanthropy

There are many steps required to truly reimagine philanthropy. Beyond just finances, these steps involve a change in mindset among donors, sector leaders and governing structures. Fortunately, most of them are under our control. So how do we implement these solutions? We've outlined many of the tools available, so let's explore concrete steps we can take to use technology to reimagine the nonprofit and philanthropic sectors.

Chapter 47

Start Using Technology

As someone who has been playing with AI tools for a few years, they have only gotten better, smarter and less 'machine-like.' Like anything, it's best to learn by doing. You can't wait for an AI playbook; your staff should be experimenting with these tools now to make their work more efficient and to save time and money. Start by asking them how they are using AI. If they aren't, point them toward the existing and free resources. Then ask them to start experimenting with simple commands such as editing a document or email. Their ability to leverage the technology will increase as they become more comfortable with it.

As we highlighted earlier, the discussion around AI currently tends to move quickly toward ethics and data security. While important, other sectors bypass these areas and go straight to the 'what if?' What if we had an app you could use to get your food delivered from any restaurant? What if we had an app you could use to get a ride somewhere? What if we had an app you could use to find a caregiver for your loved one (already exists, but a great example). The sector needs to sideline the discussion around ethics and data privacy and move to a generative discussion around 'what if?' What if all nonprofits and foundations were trained on using technology, software, hardware and AI? What if we were rewarded and had room and time to develop novel solutions to their client's needs? What if foundations allotted funds toward staff training to learn how to use technology? What if we could better serve the mother with two children struggling to find food and shelter? What if we used data to show how and what we do in real-time? What if we used data and technology to provide instant feedback to our donors?

Even with these simple, real-world applications, there exists a lack of motivation to apply technology to solve real-world problems. Funders need to change expectations of nonprofits via more room for failure and provide more funding earmarked for innovation.

Admittedly, jumping in too quickly can also be dangerous, especially if groups fail to implement proper data-security practices and fail to understand that new digital tools often come with hidden costs that can create legal vulnerabilities. In a May 2025 article from the Chronicle of Philanthropy, Jim Fruchterman, founder of Tech Matters, advises organizations to not feel like they have 'leap to the leading edge of tech. Just work your way up. Start with the cheap, generally accessible tech first,' he says. Many of your existing platforms have AI built into the system. For example, Microsoft has Copilot, your 'AI companion,' built into the top right of their browser. Their AI designer, called appropriately enough, 'Microsoft Designer,' can create images and graphics, traditionally the purview of your graphics department or outsourced marketing team, with just a few clicks.

We can also use technology as our starting point when we decide which programs to implement or which nonprofits to support. For example, when UBER was created the phone app was the primary platform for the service and the first product they developed. They didn't start with a ride sharing service and built out the infrastructure of cars and drivers, they started with the tech first. Nonprofits and foundations could adopt this same line of thinking when developing solutions; start by brainstorming which tech would be helpful to solve your problem and then consider how to adopt it and integrate it into your work. Foundations might also look for groups and nonprofits approaching their work from a 'tech first' mindset when considering who to support. See the later section on 'cross-sector' partnerships for an example.

CHAPTER 48

Increase Funding for Technology

Technology spending in the nonprofit sector typically ranges between 2-3% of revenue. Compare this with the private sector, which usually spends 3.6% and 5.8% of revenue on technology. For-profit companies are increasingly investing in revenue-generating technologies and digital transformation, while nonprofits tend to focus more on operational efficiency, mission delivery, and donor management systems. These areas require effective management, but we should weave technology into each area.

A Salesforce.com report shows that nonprofits that have embraced the use of technology to transform their work are '4x more likely to achieve mission goals.' Foundations need to start providing funding for nonprofits to purchase and use technology. The ability to use it effectively will follow.

Spending on technology implementation and growth should be a part of the costs allowed by donors and foundations. Developers and program managers who can oversee the digital transformation within a nonprofit will be critical going forward. These high-value posts may be expensive, but they are a critical part of effective service delivery that creates impact. As shown earlier, investments in capable and skilled staff lend to greater impact.

Nonprofits should be given funds to purchase technology and integrate systems and modern technological practices into their work. This should be an expectation from funders, and it should start with funders. Funders could ask questions up front to ensure nonprofits understand the importance of using technology to accomplish their mission. Funders could highlight their commitment to furthering the use of technology by asking questions in the grant application or having conversations to gauge the nonprofit's readiness for on-boarding technology during the due diligence process.

In the past, software and hardware was an afterthought in the world of funding. These costs were incorporated into administrative allocations deemed often built 'overhead.' Many funders have a ceiling on the total percentage of overhead they will allow, if they allow any at all. This policy inhibits growth and development of organizations in the sector and should be abandoned. The level of wholesale growth we need in the sector cannot fall victim to an outdated and unrealistic policy around administrative costs.

CHAPTER 49

Structural, Administrative and Tax Adjustments

By creating transparent rails showing where money is going and who is benefiting, we should pivot away from a blanket charitable tax break toward a more concise tax break that affords a greater tax benefit for donations that quickly and transparently create real-world impact.

At this point all charitable donations are taxed in different ways depending on the giving vehicle (money, art, property, for example). Donations for libraries are wonderful and can be life-changing (for me especially, as I first saw my future wife at the library where she was working as a librarian.) But the impact of these donations are not immediately realized. Building a new building takes time. Providing funds for people to eat today or students to attend college tomorrow impacts individuals and our society immediately.

We should create a tiered charitable tax deduction. The tiers would be based on the impact of the gift and reward gifts made that can be utilized immediately. One parameter should be based on the 'time to impact' or amount of time from donating to the funds actually being put to good use. In addition to the type of donation (the vehicle for giving), we should also base the amount of the deduction on whether the funds are immediately available or held in a DAF or in a foundation. Alleviating hunger immediately should be given priority and allow the donor to realize a higher tax deduction over a donation to an opera.

This tiered system would reward more donations that are transparent and have immediate impact. McKenzie Scott would be Exhibit A in the top tier of 'time to impact.' Recipients of her generosity have reported a phone call or two with a few simple questions and then receiving a large check in the mail. No long grant applications and no extended waiting periods for a response.

Furthermore, if you give your money to a DAF, you should not enjoy the same level of tax deduction as someone who give a donation to a food bank, where it will be immediately put toward the mission of the nonprofit. Conversely, the donation to the DAF, while laudable and worthy of commendation, will likely sit in the coffers of the bank or community foundation for quite some time. Nonetheless, the tax benefit to the donor is immediate in that tax year.

The point is not to reap large tax revenue for the US but to create a more transparent environment partly to encourage more donations to groups with efficient programming channels that create reliable impact. Continued opaqueness will only serve to raise more questions about the nature of the sector.

CHAPTER 50

Cross-Sector Partnerships

Leaders should step outside their comfort zone and seek out connections in the technology and entrepreneurial space. Academia is a great resource as well. A friend and mentor of mine started a meal delivery service for seniors. Volunteers would arrive each day at the central kitchen, pick up their meals and take their instructions of where to go. We partnered with engineering students to develop an app that would guide the driver though their route, similar to the uber interface. These efforts take time and there is a fair amount of upfront work that needs to happen in order to create a value proposition for an entrepreneur or student, but once you get past this initial hurdle you are usually welcomed with enthusiastic partners inspired to solve real-world problems.

CHAPTER 51

Build Capacity

Even with more funding, there is a 'technical capacity gap' in the nonprofit sector, as it lacks software engineers who are plugged into and responding to the sector's needs, creating a giant missed opportunity unfolding in real time. To build this transformation, the sector should form partnerships with technology leaders to create real-world solutions. Additional funding must come with the capacity to support new technology. We should incorporate individuals with technical skills into our work and not be so reliant on the traditional requirements for new staff.

As a former Peace Corps volunteer, it was surprising to hear technologist and Palantir CEO Alex Karp call for a 'technological peace corps.' This technological Peace Corps would presumably be deployed to apply technology to solve real-world problems. The sector should build partnerships with technology leaders to create real-world solutions. These partnerships should reach across sectors and silos. They should start with simple explanations and real-world examples of problems that need to be solved. We can build capacity at the board level to by recruiting individuals with experience in core subject areas such as technology, web design, marketing, and analytics has driven nonprofit boards to seek out new talent, including those from tech companies.

Creating relationships with experts from other sectors requires leaders to step outside their comfort zone and seek out connections in the technology and entrepreneurial space. One way to spark innovation across sectors is to host a hackathon focused on developing technological solutions to problems nonprofits face daily. At my previous employer, we hosted a hackathon to serve as a platform for local social entrepreneurs, engineers, software designers, high school and college students and local businesses to engage directly with our partner nonprofits to solve pressing organizational and programmatic challenges.

An app that connected individuals with excess food to individuals in need of food won the event. When we hosted the event in 2017, such an app was quite a novel concept. This same concept (sharing resources via an app) has since been developed by multiple organizations.

The event served the added advantage of connecting the tech sector directly with the non-profit sector, exposing the younger 'tech-first' minds to the real-world challenges the non-profit sector faces daily. Attendees had the chance to really understand the impact a tech-based solution might have on the lives of those people community-based non-profits seek to serve. The event demonstrated how the combination of technology and charitable giving fueled by ambition to solve social problems creates truly dynamic opportunities.

Academia is a great resource as well. As part of their studies, engineering and other students are often required to develop a solution for a real-world problem. A year after the hackathon, I approached an engineering professor who agreed to match his students with a local nonprofit in need of a volunteer management solution.

A friend and mentor of mine started a meal delivery service for seniors. Volunteers would arrive each day at the central kitchen, pick up their meals and take their instructions of where to go. We partnered with the engineering students to develop an app that would guide the driver through their route, similar to the uber interface. The result was a more streamlined delivery experience for the volunteer drivers. Cross-sector efforts take time and there is a fair amount of upfront work that needs to happen in order to create a value proposition for an entrepreneur or student, but once you get past this initial hurdle you are usually welcomed with enthusiastic partners inspired to solve real-world problems.

Chapter 52

Inter-Connected Computers and Networks

It can be costly and labor intensive to develop an entirely new server system and data mechanics to support projects. Consider how much Google spends on server farms and energy centers to power AI.

Utilizing the concept of de-centralization that we discussed earlier, we can invite more individuals into the philanthropic space. Some people may want to help a nonprofit but don't have the time or resources. Technology can allow individuals to connect and support a project using their own computer hardware.

A simple method is using Distributed Computer Networks that incentivize participation in research. These networks are essentially connected to computers and servers. Rewards are given back to individuals and organizations that participate in these Distributed Computing Networks. The networks utilize untapped memory and transistor power on computers and servers to create a powerful network spanning around the world that cane use used to support technology searching for important answers to medical, scientific, and mathematical problems. Members of the network are incentivized through the creation of a unique blockchain which rewards this research.

Rewards are distributed based on computer power. Anyone with a computer can become a tangible supporter of a project. These incentivize network-based giving. Further projects could be built out onto the existing slice of available network capacity to mine tokens to support researchers and nonprofits. We could also use this to power AI systems for a specific cause, say monitoring water quality in a certain region of the ocean.

Chapter 53

Start to Utilize Alternative Means of Financing

Because of the ethical considerations of the traditional finance system, and the fact that traditional forms of finance such as government grants are likely to be reduced, we should consider utilizing alternative means of financing created by emerging technology.

Traditional funding is being reduced. They are already being cut and zeroed out, with minimal or zero notice. Whether you agree or not, there are enough people who are behind efforts to reduce government spending that these trends are likely to continue. At the same time, there isn't enough funding from philanthropy and foundations to fill the gap.

New blockchain-based forms of financing are emerging. For example, within the world of finance, the blockchain is being used to invest in real-word assets (RWAs). Using financial mechanisms built onto the blockchain, individuals can now invest in assets such as companies and building or infrastructure. By creating a store of value that equates to the actual value, individuals can purchase these assets through a digital sale. Charities can leverage these developments by allowing investors to invest in their assets, such as building or equipment. This new funding can be used as an additional source of capital outside the traditional banking system, with its myriad capital requirements, interest and lending fees, thereby freeing up to the nonprofit to create financial leverage and further invest their assets into their mission. At a minimum, we should start migrating a portion of our financing onto these platforms and begin experimenting with what they offer. Currently, we own capital that sits on our books as an asset. Using emerging platforms, we can 'tokenize' our assets, allowing donors to purchase or lend capital.

CHAPTER 54

Verification

Data, inputs, ideas, concepts and intellectual property can be stolen or faked online. Inaccurate data can be generated automatically. Verification using blockchain tools built on the Ethereum network can solve this problem. First, we should incorporate blockchain layers into our work and move some or all of our work onto these platforms. Corporate entities and banks are already 'layering' payment solutions onto the blockchain. Our sector could mandate this to protect privacy, reduce donation costs and increase verification.

Verification using hardware such as retinal scanners can also be used to count program participants and ensure unduplicated counts of beneficiaries. We can also use these when we go about our daily business.

PART 11

Existing and Emerging Threats or the Risk of Apathy

CHAPTER 55

Political Threats

Originally, when the concept for this book started to emerge, the only threats I could think of revolved around the abuse of technology. Since then, for reasons discussed earlier, the political environment is increasingly a threat. One of the advantages of technology today is the ability to utilize these same technologies for verification. Verification at all levels of our sectors, the individual, program, nonprofit and foundation level, will ultimately increase accountability.

Let's start by covering the more recent political environment and consider how threats emanating from this realm could backfire on us. To re-iterate the point made earlier, the lack of accountability and transparency in the nonprofit and philanthropic sector are flaws of the system, not a built-in feature. People and actors take advantage of these conditions. Ultimately, our sector suffers.

Take for example cases where people in power with millions of dollars choose to use private foundations to channel funds to political causes. Ordinarily, funds used for political purposes must be sent through a 501c4.

The recent case of a foundation in Florida favored at the highest levels of state government is a prime example. Allegedly, a large 'donation' from a corporation was made to the foundation. Normally this is not an issue but when the corporation involved is seeking a sizable government contract, it only raises questions. In this case, the facts were slow to come out and only came to light because of leaked information, but the case was clear. Using a charitable vehicle to garner favor or influence decision making is nothing new. It will happen again. Nonetheless, this lack of transparency and antiquated information silos can knee-cap the sector. Instances such as this only further the narrative around 'these nonprofits aren't doing anything except collecting money from people and then funneling it into someone else's pocket.' Right or wrong, lies spread faster than truth. Cases such as these only chip away at the all too critical sense of trust we place in the sector.

If an elected official made the declaration tomorrow that we need to further increase taxes or levy random exorbitant fines on charitable institutions, it's hard to think that such efforts may not come to fruition, just as they have recently. The words in the first part of this book from Elon Musk, when he deemed the nonprofit sector 'a gigantic scam…like one of the biggest, maybe the biggest scams ever' still rings in my ears.

Sure, it's hard to argue that you should tax a nonprofit hospital, for example, that provides charity care to the uninsured with no other options. But what seemed impossible yesterday in some ways seems entirely possible, and sadly, all too real now.

The argument for taxing such an entity would likely not focus on the increase in tax dollars that would go to the government, but rather the argument would apply the prevailing logic of today pushed by some leaders and high-profile personalities, that being one of 'we looked at the dataset, and we can't tell where money is going…we don't know what the impact is…surely someone must be getting a kickback somewhere…and therefore it should be stopped immediately.' Yes, hospitals have patient numbers and datasets of procedures, but will that matter to the skeptics?

The risk is even greater for groups aiding marginalized or vulnerable populations who are targeted with inciteful rhetoric from the government. Some groups have had their data subpoenaed by state and federal officials.

The ability of another entity to co-op or steal our data is a fatal flaw within the systems on which philanthropy currently relies. The federal government taking steps to do this is an extreme example and hopefully not a trend.

Consider our donors and philanthropists which establish and support foundations. We hold critical information on our donors which is held by third-party servers. These corporate managed servers are largely out of our control.

Nothing positive will come from these efforts to subpoena personal data and channel funds for political reasons. Unfortunately, this trend is unlikely to change in the near-term, especially now that we live in a society that intensely pushes the narrative that universities and nonprofits are somehow at fault. The sector's data will be increasingly questioned and weaponized if we don't take steps to protect it.

Trust in the sector is one of our most valuable currencies. This trust can be eroded by individuals who take advantage of the lack of transparency afforded by our sectors. We should make efforts to solidify and strengthen the level of transparency and accountability among the sectors a priority.

CHAPTER 56

Decreasing Accountability

The way we treat data around a donor says a lot about our society. More attention is usually given at the time of the announcement. Donors often receive huge PR and attention on donations. Nonetheless, some donors won't even disclose who received the money. This is their right. Should everyone know if you buy a new luxury car? No, but there is no benefit shouldered by the taxpayer to you purchasing said car. Yet taxpayers shoulder the burden of offsetting the charitable tax benefit. This murkiness and secrecy also further perpetuate distrust in institutions and nonprofits. I would argue it even sets up the sector to be a scapegoat or easy target when tax revenue is required to offset spending increases.

Corporate donors are not left out of this debate. After a tragedy, corporations (not without good cause) often publicize the amount they are donating in response to a crisis. Wonderful stuff! But what happens after the headlines fade and our attention shifts?

In 2015, the Chan Zuckerberg initiative made sizable gifts to start schools in Palo Alto. Rightly so, there was a lot of fanfare when they announced these gifts. However, just recently in 2025, they decided to close the schools. An article which appeared shortly after the decision in the San Francisco Standard, titled 'Distraught families say Zuckerberg pulled funds from low-income school,' detailed how parents were first informed of this decision in a meeting with school officials who explained to the parents that the billionaire's foundation would move its funding elsewhere. No reason was provided for the sudden closure, nor did the foundation respond to comments with any further insight as to why the funding was pulled.

How have the students perform over the last 10 years at these schools? How many others made donations to these schools, led by the example, or donor signal, of the Chan Zuckerberg initiative? Without data that is more transparent, it's hard to know. If a donor makes a $200 million gift and writes off that amount, other taxpayers bear a partial financial burden. The donor receives the benefit, but ultimately who pays? Taxpayers offset the amount of these gifts by financing the lost proceeds to the treasury. The donor doesn't have to answer for the impact of the gift, negative or positive.

$200 million is a sizable sum. I'm not looking to 'out' someone, but it seems like a certain level of transparency could be expected by society. Increased transparency around who is donating and when would also help the sector further build trust and credibility.

In one instance provided by the article they mention the donor's use of a Donor Advised Fund (DAF) to donate. DAFs are charitable vehicles that allow donors to avoid capital gains taxes by donating stocks or other capital while claiming immediate charitable deductions on the value of the donations. These charitable accounts are held by foundations and earmarked for specific purposes which are established by the donor.

Based on the most recent data in the National Philanthropic Trust's 2024 DAF Report, charitable assets in all DAFs totaled $251.52 billion in 2023, a 9.9 percent increase from the revised 2022 total of $228.92 billion. DAFs distributed $54.77 billion to grantees in 2023. These charitable vehicles are increasingly popular, with assets growing 67% over the past four years. It is projected that assets held in DAFs will reach $2 trillion in 2026.

These vehicles warrant a closer look as we reimagine philanthropy. Robert Reich, in his 'Just Giving: Why Philanthropy is Failing Democracy and How It Can Do Better' labels DAFs as 'arcane.' Nonetheless, he acknowledges they are 'increasingly popular' and points out that foundations increasingly turn to DAFs as a means to meet the charitable payout rule. Reich, aligning this language with his 'failing democracy' title of the book, states that the use of DAFs is but a means to 'warehouse wealth until the donor decides to distribute it.' Compounded with other elements, DAFs create a 'black box' environment which serves to hold private assets and distribute them for public good, at such time as deemed appropriate by the donor.

DAF's serve a specific charitable purpose but when or where the donor money goes after they send it to a DAF is not always clear. Only the entity that holds the DAF knows when the money from a DAF was distributed, unless there is a notice or press release announcing the gift. Furthermore, the legislation establishing and governing DAFs is convoluted and difficult to navigate, unless you are a tax attorney.

While well-meaning, these vehicles tend to be opaque and do not actively contribute to the body of knowledge around where money is going, how it is being used and who is ultimately benefiting. In the end, the vehicles seemed designed for sophisticated donors with the funds to provide for a team of tax attorneys and accountants. Their role in motivating the charitable intent of the donor is questionable. See our discussion early on using AI for due diligence and I ask the question: *Is philanthropic privacy a feature, or a bug?*

CHAPTER 57

Deep Fakes

In the interview with referenced earlier between Eric Schmidt and Tim Ferriss, the topic of deep fakes came up while discussing the future of AI. Schmidt pointed out that with advancements in machine learning and artificial intelligence, 'our social world around us, within the next decade, will become impossibly confusing because there's so many actors that will want to misinform us: businesses, politicians, our opponents, for fun...or for profit.'

The notion of 'deep fakes' which are indiscernible from real videos, brings up the notion of a 'deep fake' nonprofit. If people are creating deep fake videos to influence a national election, it stands to show that someone would probably create a 'deep fake' nonprofit, replete with fundraising videos, videos of beneficiaries 'thanking' donors for their help. Anyone can put up a website, but a truly realistic platform with seemingly real donor stories is only a few clicks away. For example, American Cancer Society and United Way affiliates have been created and registered by the IRS. This raises obvious concerns for allowing the IRS to serve as gatekeeper for oversight of nonprofits.

Higher level verification of our agencies and the programs which help people every day should be a priority. Clearly, the IRS, likely just due to an overwhelming amount of information and lack of staff to keep up, is probably not the best agent to handle compliance for the sector. The private sector has fraud rates below 3 percent. Meanwhile, the public sector operates at a 20 percent fraud rate.'

The discussion around fraud at the federal government has been around for quite some time, but only now do we see broad acknowledgement and a bipartisan willingness to do something about it. The Wall Street Journal covered a recent congressional hearing in April 2025, where Haywood Talcove, CEO of LexisNexis Risk Solutions for Government said that 'for over a decade a silent war has been waged against American taxpayers and added:

'We continue to pay benefits to deceased and incarcerated individuals, direct money to bad actors flagged in the Do Not Pay list and overlook duplicate Social Security numbers by not following best practices. During the pandemic, a simple cross-check of PPP loan recipients against IRS records would have exposed massive fraud and prevented payments to transnational criminals, who sold their sauce on the dark web. To stop this, we must reclaim control of our systems not just from the criminal syndicates, but the flawed systems enabling them.'

Many of the systems the federal government currently uses to track finances and records are antiquated and rarely 'speak' to each other. It's likely the large federal bureaucracy only makes it more difficult to rectify this issue.

But if the systems we rely on use the IRS database as a primary means of compliance and rarely cross-tabulate data to verify matters, are we not heading in the same direction?

PART 12

Grantmaking Strategy to Support Networks

We talked earlier about technical and adaptive solutions. Funding one nonprofit to provide a program is a technical solution. Adaptive solutions, while they can be implemented by one nonprofit, are sometimes better created and managed by groups or networks of nonprofits. As we generally assume that a grant to a nonprofit is the entire thrust of our work, and nonprofits assume, rightly so, that their grant application can be effective in isolation from other agencies, it may be helpful for a foundation who strives to develop solutions to problems at-scale and create systems-level solutions to express a policy that supports networks of nonprofits.

While a single program implemented by a single nonprofit can change lives, a more collaborative and adaptive approach is required to scale solutions and solve problems at-scale. Networks are collaborative, adaptive solutions to adaptive problems. Expanding the focus of a foundation's grantmaking strategy beyond individual nonprofits to include groups, networks and collaboratives will result in potentially transformational work beyond the reach of a single nonprofit. Supporting networks and collaboratives also tends to be more sustainable than a grant to a nonprofit within a predetermined time frame.

This final section offers strategies for incorporating the concepts introduced in this book into our work. Below is an example of a grantmaking strategy a foundation can adopt to develop broad solutions that help us truly reimagine our work. The strategy is specific to behavioral health, but the tenets and approach can be applied to any network of collaborative.

CHAPTER 58

Grantmaking Strategy

Given the numerous organizations, entities and systems involved in supporting individuals and families seeking behavioral health care, it is essential to look beyond just the behavioral health ecosystem to identify existing efforts, untapped solutions, and potential partners. A myriad of organizations and sectors already work directly in this space. This includes the health system, law enforcement, local and state government, philanthropy and community-based organizations and the private sector. As these groups seek to collaborate, they share the common challenges such as lack of transparency, siloed approaches to funding, and capacity gaps at the staff and organizational level.

Large public sector entities charged with behavioral health services may be constrained by legislation or funding. For example, the activity and finances of the state Managing Entities are pre-defined by the legislature and strictly controlled.

Due to the interplay of private and public funding and the diverse set of systems involved in the behavioral health care system, collaboration and coordination among stakeholders and systems is essential, perhaps even more so than in other parts of the healthcare system.

With a distinct and articulated focus on creating and supporting digital connections to support networks and collaboratives, the foundation can play a key role in providing technical assistance, convening stakeholders, helping partners tell impactful stories, and bridging funding gaps when needed.

Philanthropy can fund the adoption of tools that increase coordination between call centers, crisis response teams, stabilization centers, and other facilities for higher level support. The foundation can support planning, alignment, and learning among key stakeholders to adopt and utilize connecting technology. If collaborative efforts are already underway, the foundation can bolster their capacity through designated staffing or by hosting educational meetings focusing on technology.

If not yet established, the foundation can use their influence and reputational capital to create opportunities in partnership with local experts for key stakeholders to come together to assess the existing landscape, support planning, and implement technological reform. Lastly, the foundation can actively support the projects and opportunities that stem from this collaboration.

Strategic Objectives:

- Convene foundations in the state of Florida already working in mental and behavioral health to develop a collaborative network of philanthropic leaders dedicated to mental health access.
- Support mission-aligned organizations working collectively to develop partnerships or grow networks of individuals, organizations and other partners which increase awareness of available services and create synergies across groups.
- Provide general operating support to existing and emerging collaboratives and networks.
- Support groups which include a focus on at-risk populations lacking ready awareness and/or access to behavioral Health Services such as; youth, rural areas and Pre- and Post-Natal African American Mothers.
- Convene and connect managing entities, hospitals, health systems and the private sector to help address gaps, and ensure their resources drive meaningful change.
- Create collaboratives and commit to developing and sustaining their efforts over the long-term.

CHAPTER 59

Grantmaking Strategy to Reimagine Technology

Strategic Opportunity: Technology

Here is another grantmaking strategy specifically tailored to further the use of technology. Again, specific to behavioral health but the approach can be applied to any number of sectors.

Technology, telehealth and the use of digital health solutions can increase access to behavioral health and crisis care, especially in rural areas. Technology can also advance the implementation of the integrated care model by supporting workflows, enhanced care coordination and patient communication. It is a useful tool in connecting across systems and among organizations and providers.

According to the National Institute of Mental Health, 'technology has opened a new frontier in mental health care and data collection. Mobile devices like cell phones, smartphones, and tablets are giving the public, health care providers, and researchers new ways to access help, monitor progress, and increase understanding of mental well-being...excitement about the huge range of opportunities technology offers for mental health treatment has led to a burst of development. Thousands of mental health apps are available in iTunes and Android app stores, and the number is growing every year. However, this new technology frontier includes a lot of uncertainty.'

Technology and digital solutions generally require an up-front cost to obtain the service and equipment. Inevitably, these new capital costs introduce requirements for staff training and workflow integration. The foundation can help offset the expense related to obtaining technology and training staff in how to use the technology. In addition, patient-facing technology and remote monitoring technology often require training for the patient. This is another cost-center that the foundation could help offset by providing space and time for patient training.

The foundation can take an active role in promoting awareness, development and utilization of technology. This strategy would require partnerships with the private sector and academia. Serving as a connector and convenor, the foundation could promote cross-sector communication and development of useful technological tools to solve many of the challenges inherent in accessing and delivering behavioral health services in Florida.

Many nonprofits find it difficult to obtain the required software and training to integrate technology into their daily workflow. In addition, staff and patients may need training on how to use new technology. Lastly, the amount of information available and flowing within organizations tends to be engineering in a way that results in siloed and separated informational systems. The ease of creating new software solutions along with the integration of AI into our daily workflow represents a unique opportunity.

Outside of the private sector, this area is still an area mostly untapped. Leveraging new partnerships would put the foundation at the forefront of new developments.

Objectives:

- Provide education and awareness on existing resources to individuals, nonprofits and nonprofit staff.
- Support the acquisition of software and necessary staff training to incorporate technology into existing organizations and practices.
- Support training for patients and end-users on patient-facing technology and remote monitoring technology.
- Directly engage post-secondary students in the development or testing of new or existing technology.

- Build partnerships with startup incubators, high schools and colleges with computer science, engineering and health care schools.
- Connect academic research and engineering departments with end-users (individuals and nonprofits) to combine skills and experience to create meaningful technological solutions which can address a wide range of behavioral health challenges.
- Create greater understanding of the existing financing streams for the behavioral health system by creating a data-sharing network which connects siloed organizations and entities.
- Foster an environment among partners which promotes and supports the integration of technology into partner organizations.

Epilogue

John D. Rockefeller first conceived of what we now know as a foundation in 1909. Instead of being met with appreciation for his generosity, he was met with deep skepticism. Delays and this skepticism prolonged the effort to get the movement off the ground until 1913.

We have the tools to reimagine giving, we just need the courage. Many of us know the great benefits of philanthropy and appreciate the role of the foundation in the betterment of society, but we should never take the work for granted. We should hold it dear and protect it for future generations by taking ownership of our work and our sector.

Bibliography

Chapter 1:

July 26, 2025. *The economics of superintelligence: If Silicon Valley's predictions are even close to being accurate, expect unprecedented upheaval.* The Economist.

Ferriss, T. Tim Ferriss. October 27, 2021.*Eric Schmidt — The Promises and Perils of AI, the Future of Warfare, Profound Revolutions on the Horizon, and Exploring the Meaning of Life* (#541). The Tim Ferriss Show Transcripts.

Chapter 2:

Camarena, J. November 29, 2022. *GlassPockets at sunset: reflections from working at the intersection of power and light.* November 29, 2022. Candid.org.

Bender, M. Blinder, A. Schmidt, M. July 28, 2025. *Harvard Is Said to Be Open to Spending Up to $500 Million to Resolve Trump Dispute.* New York Times.

U.S. Trust. July 2018. *The U.S. Trust Study of the Philanthropic Conversation: Understanding Advisor Approaches & Client Expectations Conducted in partnership with The Philanthropic Initiative.*

Rogan, J. Feb 28, 2025. Elon Musk on Joe Rogan Experience Podcast #2281. Joe Rogan Experience #2281.

Part 3:

Putnam-Walkerly, K. 2020. *Delusional Altruism: Why Philanthropists Fail To Achieve Change and What They Can Do To Transform Giving.* Wiley.

Chapter 7:

Lozier, T. 2024. *Future of Philanthropy:3 Key Takeaways from TAG's State of Philanthropy Tech Report 2022 (Updated for 2024)* Fluxx.

Chapter 8:

Salesforce. October 27, 2022. *Just 12% Of Nonprofits Are "Digitally Mature" — And They're 4X More Likely to Achieve Mission Goals.*

Chapter 10:

Pollatta, D. January 2020. *The Flat Org Chart. We're Not Organized to Solve Problems in Our Communities. And Our Dreams Aren't Big Enough to Change That.* CThings.

Miller, C. September 28, 2022. If Foundations Want to Encourage Transparency, They Should Look in the Mirror. The Chronicle of Philanthropy.

2025 *Statistics on US Generosity.* Philanthropy Almanac. The Philanthropy Roundtable.

Brock, A. Buteau, E. Gopal, R. 2013. *Foundation Transparency: What Nonprofits Want.* The Center for Effective Philanthropy, Inc

Part 4:

Davis, S. 2021. *Solving the Giving Pledge: How to Finance Social and Environmental Challenges Using Venture Philanthropy at Scale.* Palgrave MacMillan.

Chapter 38:

Ward, K. December 21, 2021. *How Black Communities Become "Sacrifice Zones" for Industrial Air Pollution.* Mountain State Spotlight.

Chapter 44:

Baer, J. Heeb, G. May 22, 2025. *Big Banks Explore Venturing Into Crypto World Together With Joint Stablecoin.* The Wall Street Journal.

Freeman, J. February 12, 2025. *Trillion-Dollar Fraud. A rare bipartisan consensus in Washington acknowledges that the amount of theft is massive.* The Wall Street Journal.

Chapter 47:

Dickey, M. May 31, 2025. '*The Growing Nonprofit Digital Divide: An Exclusive Chronicle Survey.*' Chronicle of Philanthropy.

Scheid, Randy. 2017. *Code4GoodPBC: Bridging the Gap between Innovation and Philanthropy.* Computer 50(7): 29-31.

Chapter 51:

Karp, A. Zamiska, N. February 18, 2025. *The Technological Republic: Hard Power, Soft Belief, and the Future of the West.* Crown Currency.

Chapter 56:

Leahy, G. Shugerman, E. April 23, 2025. *Distraught families say Zuckerberg pulled funds from low-income school.* San Francisco Standard.